Lecture Notes in Networks and Systems 1213

The series "Lecture Notes in Networks and Systems" publishes the latest developments in Networks and Systems—quickly, informally and with high quality. Original research reported in proceedings and post-proceedings represents the core of LNNS.

Volumes published in LNNS embrace all aspects and subfields of, as well as new challenges in, Networks and Systems.

The series contains proceedings and edited volumes in systems and networks, spanning the areas of Cyber-Physical Systems, Autonomous Systems, Sensor Networks, Control Systems, Energy Systems, Automotive Systems, Biological Systems, Vehicular Networking and Connected Vehicles, Aerospace Systems, Automation, Manufacturing, Smart Grids, Nonlinear Systems, Power Systems, Robotics, Social Systems, Economic Systems and other. Of particular value to both the contributors and the readership are the short publication timeframe and the world-wide distribution and exposure which enable both a wide and rapid dissemination of research output.

The series covers the theory, applications, and perspectives on the state of the art and future developments relevant to systems and networks, decision making, control, complex processes and related areas, as embedded in the fields of interdisciplinary and applied sciences, engineering, computer science, physics, economics, social, and life sciences, as well as the paradigms and methodologies behind them.

Indexed by SCOPUS, EI Compendex, INSPEC, WTI Frankfurt eG, zbMATH, SCImago.

All books published in the series are submitted for consideration in Web of Science.

For proposals from Asia please contact Aninda Bose (aninda.bose@springer.com).

Phil Legg · Natalie Coull · Charles Clarke
Editors

Advances in Teaching and Learning for Cyber Security Education

Springer

Editors
Phil Legg
School of Computing and Creative Technologies
University of the West of England
Bristol, UK

Natalie Coull
School of Design and Informatics
Abertay University
Dundee, UK

Charles Clarke
University of Roehampton
London, UK

ISSN 2367-3370 ISSN 2367-3389 (electronic)
Lecture Notes in Networks and Systems
ISBN 978-3-031-77523-9 ISBN 978-3-031-77524-6 (eBook)
https://doi.org/10.1007/978-3-031-77524-6

This Springer imprint is published by the registered company Springer Nature Switzerland AG
The registered company address is: Gewerbestrasse 11, 6330 Cham, Switzerland

Preface

Cyber security is recognised as a necessary requirement for our modern society to function, to ensure that citizens can conduct their professional and personal lives alongside the advances that technology brings. The UK National Cyber Security Centre (NCSC) sets out to ensure that the UK remains the safest place to live and work online, and as part of this pledge, the NCSC works closely with academia to acknowledge and inform cyber security education within the UK Higher Education sector. The NCSC actively promotes degree certification in recognition of high-quality cyber security courses, as well as recognition of Academic Centres of Excellence in both Cyber Security Education (CSE) and Cyber Security Research (CSR). As a growing community, these institutions represent the exemplar of good cyber security practice beyond their institutions, to help upskill and contribute towards the wider cyber security ecosystem.

CSE-Connect has evolved as a collaborative effort to bring together academia, industry and government to work together to solve these cyber security educational challenges, supported by the NCSC and led by academics that share a collective vision to bring together a community response to the current challenges. Collaboration is key, and promoting opportunities for how this can be achieved is vital. As a community group, we established in 2019 as a UK chapter of Colloquium of Information Systems Security Education (CISSE), a well-established community group in the USA that has helped to inform how we can establish and grow and promote our efforts not only within the UK but across the globe. Our first conference was held in July 2019 at the University of the West of England, where we invited industry, government and academia to be part of this community movement and hear from speakers who are actively leading collaborative efforts in this domain. Following the global pandemic, subsequent conferences have been held at the University of Roehampton in 2022 and Abertay University in 2023.

This book is a showcase of our 4th Annual Conference in Advances in Teaching and Learning for Cyber Security Education and represents how our community continues to grow and collaborate to achieve fantastic work. We are delighted to have received a number of contributions from authors across the UK Higher Education sector, to share and collaborate on best practices, innovative teaching methods and new frameworks for how we educate and assess within the realm of cyber security. We hope that the readership of this book can benefit from the many practical hands-on activities and resources that are described by the authors, as our intention is to help further disseminate the excellent work of our academic colleagues across the Higher Education sector, and to bring together a collaborative response to how we ensure that our students have the best experience in learning about cyber security, and that they benefit from the wider community and not only their individual institutions.

We would like to thank the authors of the contributed works in this book, as well as all the attendees of the 4th Annual Conference in Advances in Teaching and Learning for Cyber Security Education that was hosted at the University of the West of England on 2nd July 2024. The conference brought together over 170 attendees from academia, industry and government and served as an excellent opportunity to build new collaborations as well as embrace existing linkages and to help strengthen the collaborative and combined forces that our community has to offer.

September 2024

Phil Legg
Natalie Coull
Charles Clarke

Contents

List of Contributors

Aida Abzhaparova University of the West of England, Bristol, UK

Drew Bassett Department of Computer Science, Swansea University, Wales, UK

Matthew Bowden University of Birmingham, Birmingham, UK

Vesna Brujic-Okretic Kingston University London, Thames, UK

Michelle Butler Queen's University Belfast, Belfast, UK

Ian Caple University of the West of England, Bristol, UK

Arunima Chaudhary Cardiff University, Cardiff, UK

Tom Chothia University of Birmingham, Birmingham, UK

Anna Clee University of Birmingham, Birmingham, UK

Sam Collins University of Birmingham, Birmingham, UK

Walter Colombo Cardiff University, Cardiff, UK

Rosanne English University of Strathclyde, Glasgow, UK

Sophie Fenn Pate's Grammar School, Cheltenham, UK

Jacqueline Henes University of Birmingham, Birmingham, UK

Amir Javed Cardiff University, Cardiff, UK

Beth Jenkins University of South Wales, Newport, UK

Harjinder Lallie University of Warwick, Coventry, UK

Phil Legg University of the West of England, Bristol, UK

Joseph Maguire University of Glasgow, Glasgow, UK

Fernando Alva Manchego Cardiff University, Cardiff, UK

Rachael Medhurst University of South Wales, Newport, UK

Nicholas Micallef Department of Computer Science, Swansea University, Wales, UK

Alan Mills University of the West of England, Bristol, UK

Faron Moller Department of Computer Science, Swansea University, Swansea, UK

Berndt Müller Department of Computer Science, Swansea University, Wales, UK

Jacques Ophoff Abertay University, Dundee, UK;
University of Cape Town, Cape Town, South Africa

David Oswald University of Birmingham, Birmingham, UK

Carole Parsons Queen's University Belfast, Belfast, UK

Eckhard Pfluegel Kingston University London, Thames, UK

David Pickard Kingston University London, Thames, UK

Suzanne Prior Abertay University, Dundee, UK

Joanne Ralph CyberFirst, Swansea, UK

Z. Cliffe Schreuders Leeds Beckett University, Leeds, UK

Sandra Scott-Hayward Queen's University Belfast, Belfast, UK

Thomas Shaw Leeds Beckett University, Leeds, UK

M. Rehan Usman Kingston University London, Thames, UK

Megan Venn-Wycherley Department of Computer Science, Swansea University, Swansea, UK

Adriana Wilde University of Southampton, Southampton, UK

Exploring the Safe Integration of Generative AI in Cybersecurity Education Addressing Challenges in Transparency, Accuracy, and Security

Arunima Chaudhary(✉), Amir Javed, Walter Colombo, and Fernando Alva Manchego

Cardiff University, Cardiff, UK
ChaudharyA4@cardiff.ac.uk

Abstract. The educational landscape has undergone a profound transformation in the wake of the pandemic, necessitating a swift pivot to digital pedagogy. The surge in popularity of Generative AI (GenAI) tools, exemplified by a million subscribers within the first week of its release, underscores their inevitable integration into teaching. It is clear that the genie is out of the bottle, and AI is now so deeply integrated into the learning experience that it would be counterproductive to resist the change. Meanwhile, in the age of emerging technologies and cyber threats, Cybersecurity Education has become of paramount importance. This paper explores the integration of GenAI within cybersecurity education, aiming to bridge existing gaps and enhance learning outcomes. By addressing the themes of accuracy, transparency and security in GenAI tools, the review provides insights into the ethical and effective use of AI in education, aiming to prepare students to face evolving cyber threats. The critical question explored is: Why can't Generative AI be effectively incorporated as an assistive tool for both educators and students in Cybersecurity Education? The paper highlights the potential of AI-driven educational tools to transform cybersecurity training, ensuring that students are well-equipped with the necessary knowledge and skills to navigate and protect the digital landscape.

Keywords: Cybersecurity Education · Generative AI · Cybersecurity · Cybersecurity challenges

1 Introduction

The advancement of education through digital technology has evolved rapidly in recent years. Among these developments, Artificial Intelligence (AI) has proven to be particularly transformative in academia, especially with the emergence of Generative AI tools. (Meli, Taouki and Pantazatos, 2024) claims that these tools have the potential to reshape critical aspects of the teaching and learning landscape.

In addition to general educational applications, this review focuses on the role of Generative AI in cybersecurity education and the challenges, limitations and gaps of

P. Legg et al. (Eds.): CSE 2024, LNNS 1213, pp. 1–21, 2024.
https://doi.org/10.1007/978-3-031-77524-6_1

the same. As digital technologies advance, the need for robust cybersecurity education becomes increasingly critical, especially for students who are early adopters of new technologies, online content, or those aspiring to a career in this field (Catal et al., 2023). Generative AI tools offer promising opportunities to enhance the teaching and learning of various cybersecurity concepts by providing real-time data analysis, creating real-world scenarios, simulating cyber threats, and assisting in the identification of threats and vulnerabilities but here hasn't been much research done to combine these two fields and come up with a solution to use Generative AI tools in cybersecurity education(Michael, Abbas and Roussos, 2023).

This review synthesizes existing research on the integration of Generative AI in education particularly Cybersecurity education, focusing on identifying gaps in current work, exploring the benefits and challenges of this integration, and providing insights into the concerns of using AI in educational settings. The themes explored include transparency, accuracy, and security. By delving into these themes, this review aims to contribute to the understanding of how Generative AI can be used in educational settings to enhance student experiences with valid, safe, and secure information, ultimately improving learning outcomes and addressing gaps in cybersecurity education.

Due to their robust natural language processing (NLP) capabilities, most Generative AI tools can process and respond to prompts in a human-like manner, providing solutions, breaking down complex topics, and offering innovative and personalized educational resources (Zhu et al., 2023). These tools assist teachers in creating interactive quizzes, providing feedback on coursework, automating tasks, and developing customized learning materials for students (Trust, Whalen and Mouza, no date). However, the training data for these tools is sourced from various repositories that may contain incomplete or biased information, leading to inaccuracies and misinformation often reflected in their responses (Brown et al., 2020).

The integration of Generative AI in education raises critical concerns regarding plagiarism and reliability. A survey by, ('Student perceptions of generative AI report', no date) ran nine series of in-person discussion forums with over 200 students across several colleges and universities show that issues such as fabricated content and inaccurate results are significant, and students have expressed concerns about copyright from two angles: "Who owns the work co-created with generative AI tools?" and "How can they ensure that they are not inadvertently plagiarizing someone else's work?" (Munn, Magee and Arora, 2023).

The review on Exploring the safe integration of Generative AI in Cybersecurity Education centers its attention on the secure and effective usage of Generative AI tools in Cybersecurity Education. The existing literature on Generative AI and Cybersecurity highlights its contribution to respective fields, yet their combination in educational context could lead to further advancements in filling the existing gaps of lack of cybersecurity in all levels of education by the innovation of Generative AI tools in digital pedagogy.

The research questions addressed in this review are:

1.1 Research Questions

RQ1: Why is cybersecurity education important for students in the context of rapidly advancing digital technologies?
RQ2: What are the gaps in cybersecurity education?
RQ3: What are the benefits and challenges of safely incorporating Generative AI in cybersecurity education?

1.2 Contributions

- This paper explores the benefits of Generative AI for both teachers and students, highlighting how these tools can enhance learning experiences and provide personalized support.
- It identifies and explores the existing gaps in cybersecurity education.
- It examines the challenges of incorporating Generative AI in education, specifically in cybersecurity education, and provides insights into institutional resistance.
- The study narrows down the challenges of Generative AI tools in education into three main themes: transparency, accuracy, and security, and examines the existing work proposed to overcome these challenges.

2 Methodology

For this study, a scoping review was conducted to ensure comprehensive coverage of the most relevant literature at the intersection of cybersecurity education and generative AI. This approach was chosen to review and identify knowledge gaps, map the scope of existing literature, clarify key concepts, and examine the conduct of research in this field (Munn et al., 2018). Unlike systematic reviews, scoping reviews are broader in scope and typically address exploratory research questions. This will make them a valuable precursor for recommending a more detailed systematic review in future research.

The PRISMA-ScR (PRISMA Extension for Scoping Reviews) guidelines were primarily followed to select relevant papers. Although there already exists a PRISMA checklist for systematic reviews, a revised one for scoping reviews was published which is followed in this paper to ensure the specific type of knowledge synthesis. While not every step of PRISMA-ScR was strictly adhered to, the framework was chosen to ensure a clear understanding of relevant terminology, core concepts, and key items to report in scoping reviews. The PRISMA method promoted transparent reporting of the review process, thereby enhancing the quality and reproducibility of the research (Tricco et al., 2018).

The search process to select the relevant papers was a hybrid search strategy of Automatic search, Manual search and Snowballing (Wohlin et al., 2022).

2.1 Automatic Search

The initial search was conducted on Google Scholar using the primary keywords "Cybersecurity Education" and "Generative AI," with the publication date restricted to "since 2020." These keywords were selected as they form the foundation of the research. The time constraint was applied to ensure the relevance of the papers, given the emerging nature of the topic. The selection process was based on the eligibility criteria from the PRISMA checklist to ensure the inclusion of up-to-date literature.

The year 2020 was specifically chosen because a significant aspect of the research is the rise of Generative AI following the pandemic. This initial search yielded 14,700 results. To further refine the search and cover all potential papers, additional keywords such as "ChatGPT" and "GenAI" were included. This expanded the search to 14,800 results.

2.2 Manual Search

Following the extensive results from the automatic search, further filtration was conducted manually by examining specific libraries such as IEEE Xplore, ScienceDirect, ACM, Arxiv, Springer, and journals like CISSE, as well as published papers on OpenAI's official site. These libraries and journals ensured trustworthy and reliable papers, leading to credible quality and up to date research. The initial selection of papers was based on their titles. Relevant papers were then reviewed and filtered based on their abstracts and introductions. Subsequently, the snowballing method described by (Wohlin, 2014) was employed. This involved using the reference lists of selected papers to identify additional sources. Forward snowballing was used to identify new papers citing the selected ones, while backward snowballing was applied to find new papers referenced by the selected ones. During this process, papers that did not meet the following criteria were eliminated:

- Papers in other language than English.
- Publication year older than 2020.
- Papers where the full text is not available online.
- Papers should be a full research paper and not a poster or a presentation.
- The paper's primary focus on Cybersecurity, Cybersecurity education and Generative AI or a combination of any two.

Through continuous forward and backward snowballing, a final set of approximately 80 papers were selected for the comprehensive review.

The flowchart, Fig. 1 illustrates the methodology for conducting the literature review on the integration of Generative AI in cybersecurity education.

Purple box: These boxes represent the major stages and decision points in the process, starting from the initial search keyword list to the final selection of literature review candidate papers.

Yellow box: These boxes denote the specific actions taken during the literature review process. This includes the automatic search on Google Scholar, manual search by title, filtering by abstract, and the snowballing technique to identify additional relevant papers.

The process involves both automated and manual search strategies, followed by filtering and refinement to ensure a comprehensive and relevant set of literature for review.

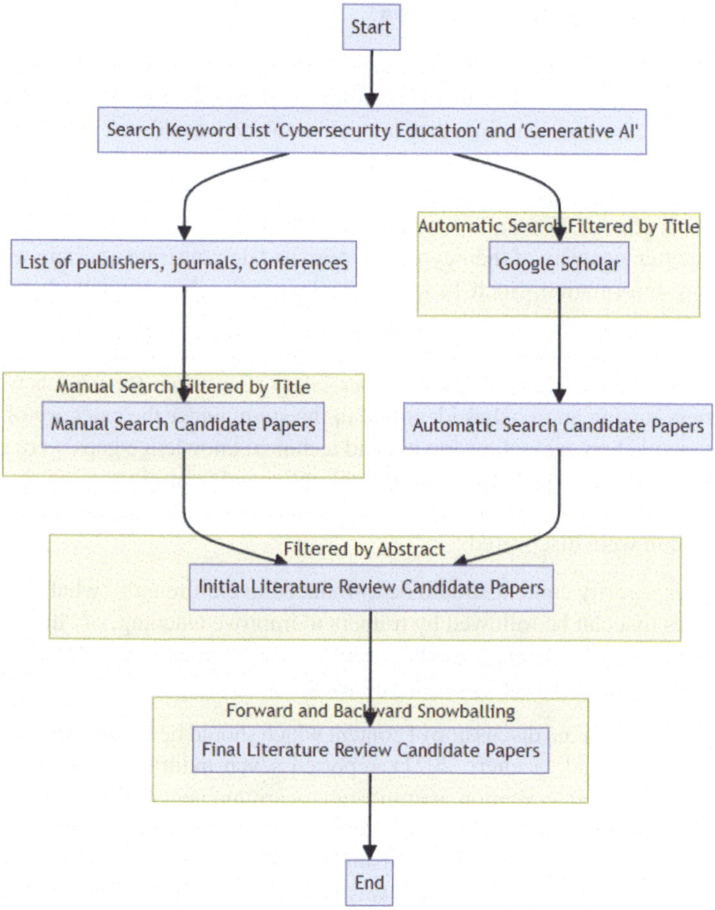

Fig. 1. Flowchart of Research Methodology

3 Related Work

3.1 Importance of Cybersecurity Education

Cyber security is a critical aspect of modern information technology as it helps protect sensitive data, programs, files, and software from unauthorized access and potential harm from cyber attackers. With the growing use of online resources and the rapid advancement of technology, cyber security remains a top concern for individuals, businesses, and governments (Kumar et al., 2023). Thus, to address the ever-growing threat landscape in cyberspace, cybersecurity education becomes of paramount importance. (A. Ghaleb et al., 2020). These skills become even more critical for younger people as they are the early adopters of new technologies and online content.('Children and parents: media use and attitudes report 2022', no date).

3.2 Need for Cybersecurity Education

(Hourcade, 2015) states that children are growing up in a world where more of their lives are digitally recorded. Whether it is online gaming, social media, browsing on the internet, or asynchronous learning, they are vulnerable to leave their cyber footprint (the profile that they leave online). (Pencheva, Hallett and Rashid, 2020) conducted interactive workshops across the UK to identify drivers and barriers in introducing cybersecurity into secondary school education and found out that during discussion, it was suggested that students are often unaware of their cyber footprint and demonstrate a certain willingness to give away information. Such lack of knowledge of cyber etiquette has led some students to post illegal content online, such as having social media accounts at an age as young as 9 years or posting either their personal information or inappropriate images of themselves or their classmates. Making cyber-safety curriculum for schools was one of the awareness-raising approaches identified in the study under the category of training. In the same workshop, lack of awareness and technical knowledge gaps were identified as a significant factor contributing to students' online vulnerability.

According to (Sag, 2023) thematic analysis, three main themes regarding cybersecurity education were discovered:

- what cyber security content should be introduced to children, or "what to teach"
- approaches that can be followed by trainers to improve teaching, or "how to teach"
- parties who are responsible for cyber security education amongst children and their competence in this area, or "who should teach".

However, there was no discussion of content which should be covered in a curriculum. (Quayyum, Cruzes and Jaccheri, 2021), reported seven main methods (i.e., training, game-based learning, intervention, gamification, warning, negotiation of cyber security within the family home, and mobile app) for teaching. Making cyber-safety curriculum for schools was one of the awareness-raising approaches identified in the study under the category of training. However, there was no discussion of content which should be covered in a curriculum.

(Kumar et al., 2023) highlights significant gaps in cybersecurity education. The paper emphasizes that the constantly evolving nature of cybersecurity challenges means many individuals are unaware of the latest threats and vulnerabilities, such as social engineering attacks, phishing scams, and ransomware. This underscores the necessity for ongoing training and education to keep individuals updated on current threats and effective mitigation strategies. (Kumar et al., 2023) also points out that a strong technical foundation is crucial for cybersecurity, yet many individuals lack the necessary technical skills and knowledge. For example, they may not understand how firewalls, antivirus software, or intrusion detection systems function, making it difficult to identify and prevent attacks effectively. There is an emphasis on the importance of comprehensive education and technical training to address these critical gaps in cybersecurity.

(Grover, Broll and Babb, 2023) highlight that cybersecurity education is crucial for learners, emphasizing that it is impossible to have meaningful cybersecurity discussions today without involving AI and ML. Integrating these two rapidly evolving subfields of computing into cybersecurity education makes the learning experience relevant to today's technological landscape. Offering engaging learning experiences in secondary schools fosters career aspirations and interest in these fields, making these educational efforts productive and beneficial for the future workforce.

3.3 Generative AI in Education

In a scoping literature review done by (Law, 2024) it is showcased that Generative AI models provide a platform for personalized learning, real-time response, improved language learning outcomes, and enhanced learning experiences and autonomy. Learners can set objectives plan and regulate their learning of a specific topic via tailored language learning assistance. (Zhu et al., 2023).

(Haleem, Javaid and Singh, 2022) mention that ChatGPT differs from previous AI models, as it can write software in many languages, debug code, break down a complicated subject into manageable chunks, prepare for interviews, and draft essays. ChatGPT simplifies such processes and even provides the result, much as how one might research similar subjects online.

3.3.1 For Students

(Chiu et al., 2023) classifies AI in education into four main roles: (i) assigning tasks based on individual competence, (ii) providing human–machine conversations, (iii) analyzing student work for feedback, and (iv) increasing adaptability and interactivity in digital learning materials. It can be used by students to provide personalized feedback and act as an assistant tutor to learn according to their needs and progress, resulting in improved learning outcomes. Students can interact with these tools in a conversational manner and prove to be beneficial to provide support, especially for students who are non-English speakers and want a customized learning experience in their native languages (Baidoo-Anu and Ansah, no date).

Generative AI tools like chatbots and Intelligent Tutoring Systems act as digital personal tutors that can respond to prompts and assist students in giving them real-time answers in a human-like manner. These systems use AI tools like Machine Learning (ML) and NLP to interact with students in a conversational style, responding to inquiries, gauging their comprehension, and providing customized explanations (Sahu, no date). (Xu and Ouyang, 2022) highlights that AI has enhanced learning environments through personalization and emphasizes that this technology has the potential to forge adaptive educational pathways.

3.3.2 For Teachers

Similarly, (Hsu and Ching, 2023), categorizes Generative AI's use in education by teachers into five areas: 1) provide support with teaching; 2) provide support with student assessment; 3) help support student learning; 4) offer suggestions for improving teaching; 5) support teacher-parent and teacher-student communication.

Thus, research consistently shows how OpenAI tools can generate interactive quizzes and learning materials, creating more engaging and effective educational sessions. These tools significantly enhance learning experiences by making content more engaging and promoting quick learning. For teachers, this technology offers a valuable opportunity to save time, allowing them to focus on other areas of development and delivery. By utilizing these tools in the same way students do, as assistants for generating interactive materials and ideas, teachers can better align with the shift towards digital learning. This approach not only supports the creation of fun and engaging educational content but also adapts to the evolving learning preferences of students.

3.4 Challenges of Generative AI in Education

(Thurzo et al., 2023) highlights the growing concern of the safe use of AI and states that fact-checking is required to integrate and make full use of large language models in classroom styles as it includes the potential for bias in the output, the requirement of continual human monitoring, and the potential for misuse. These issues arise as a cause for concern because the Generative AI tools like ChatGPT are effectively just text generating machines, which do not grasp the information they generate, nor do they judge whether it is accurate or relevant. (Mhlanga, D., 2023).

(Tlili et al., 2023) conducted a study, 'What are the concerns of using chatbots, specifically ChatGPT, in education?', which is a qualitative case study, benefitting from an instrumental case study research design. The study was divided into three stages:

- social network analysis of tweets
- content analysis of interviews and
- investigation of user experiences.

From the above three stages, the first stage, i.e., social network analysis of tweets, (Tlili et al., 2023) found that however the positive sentiments outweigh the negative sentiments, the negative sentiments demonstrated a deeper and critical thinking over the positive ones.

For the second stage, i.e., content analysis of interviews, there was a positive response by majority of the participants, who recognized ChatGPT to be efficient in providing a comprehensive understanding of various topics, demonstrating a paradigm shift in the conventional approaches of education. Alternatively, there were some concerns about these tools affecting the innovative and critical thinking of the users, leading to easy-to-get solutions and being dependent on them.

In terms of response quality and usefulness, participants generally expressed appreciation for the dialogue quality, specify and relevant information provided by ChatGPT, highlighting its ability to engage in coherent and contextually appropriate conversations. However, a critical finding emerged regarding concerns about the accuracy and misinformation generated by the tool. Specifically, participants noted the presence of errors and gaps in the responses, underscoring the significant need to improve the output quality of ChatGPT. It was noted that alternative answers would sometimes contradict previous answers of the same topic and had a limitation of providing contextual information.

A particularly notable example involved the generation of incorrect code by ChatGPT, which failed to execute correctly in programming software. This instance underscores the potential risks and limitations associated with relying on generative AI tools for precise and technical tasks. Despite these issues, some participants viewed the tool's shortcomings in a more positive light. They praised ChatGPT as an effective assistant tool for constructing knowledge, emphasizing that even with its errors, it can serve as a valuable aid in the learning process. This dual perspective highlights the nuanced role that generative AI can play in education and knowledge acquisition, where its strengths in dialogue and support must be balanced against the need for accuracy and reliability (Tlili et al., 2023).

In the same study (Tlili et al., 2023), when analysing personality and emotions, it was observed that some participants had developed an emotional attachment to ChatGPT, considering it a personal tutor. However, their trust and confidence were significantly undermined when they discovered that some of the information generated by the tool was incorrect. This realization led to feelings of frustration and a loss of trust among the participants. This finding underscores the potential negative impact of misinformation produced by generative AI tools, particularly when users have formed a personal attachment with the technology.

3.5 Concerns of Using Generative AI by Teachers

There has been a significant resistance to the use of Generative AI tools in schools and universities in the wake of its potential risk, challenges and AI-assisted cheating. Numerous educational institutions like the New York City Department of Education have blocked chatbots from their school networks across the districts, following the concerns of plagiarism in student's works. (Zhu et al., 2023).

In Cardiff University's 2nd AI in Education Conference (*Artificial Intelligence (AI) in education conference*, 2024), a prominent discussion emerged among academics from various schools, revealing a clear divide on the integration of Generative AI tools in university settings. The audience was split into two distinct groups.

The first group was curious and eager to integrate Generative AI into their schools, aiming to promote a safe integration for students who already use these tools for learning. They believed that leveraging AI could enhance educational experiences and keep pace with technological advancements.

The second group, however, opposed this official integration for three main reasons:

- Impact on critical thinking and comprehension: They argued that reliance on AI tools might lead to a decline in students' critical thinking, self-analysis, and comprehension of topics.
- Increased workload for teachers: There was concern that integrating AI would increase teachers' workload. This would include the need for additional training, raising awareness, and incorporating an extra step in the assessment process.
- Preservation of traditional Learning practices and slow learning: The third concern was the need to protect the concept of slow and comprehensive learning practices, which lead to a deeper understanding of topics. They advocated for sticking to traditional teaching methods, emphasizing thorough and reflective learning processes.

This division highlights the ongoing debate about the role of Generative AI in education, reflecting both the enthusiasm for innovative tools and the caution towards preserving educational integrity and quality.

4 Discussion

4.1 Generative AI in Cybersecurity Education

(McIntosh et al., 2023) states that recent advancements in Artificial Intelligence, hold significant promise for improving the efficiency and effectiveness of Governance, risk Management, and Compliance (GRC) processes in cybersecurity policy development and decision-making. To which (Nowrozy and Jam, 2024) agrees that this underscores the importance of tertiary education providers incorporating generative AI into cybersecurity degree curricula as they are expected to play a vital role in workforce development. Tertiary institutions are frequently expected to play a vital role in workforce development. They do this by equipping graduates with the essential knowledge and skills needed for their careers, acting as a bridge between academic learning and practical application, and assisting students in transitioning from the classroom to the workplace. (Nowrozy and Jam, 2024) focuses on cybersecurity education as the primary discipline of interest, among others as cybersecurity industry is both highly sensitive to the adoption of AI and has already demonstrated significant adoption of AI technologies. In light of this, institutions should assess the compatibility of their current cybersecurity curricula with the rise of GenAI, and proactively evaluate and update their curricula to address such changes, to ensure that their graduates are equipped with the most relevant and up-to-date knowledge and skills to remain competitive in the job market. In this intensive survey done by (Nowrozy and Jam, 2024), they highlight that universities must align their cybersecurity curricula with industry demands and expedite the integration of Generative AI technologies, such as GPT, to prepare graduates for the evolving landscape. This addresses the gap of this research to integrate Generative AI in Cybersecurity education and use it as a tool for knowledge dissemination.

(Wei-Kocsis et al., 2024) mention that although efforts have been made to bridge the education gap that throttles aspirations in the advance of AI and intensify the shortage problem in the cybersecurity workforce, there still contain essential challenges for effectively educating students on the interaction of AI and cybersecurity. They have stated the following examples: 1) due to the emerging and growing features of AI technologies, the integration of AI and cybersecurity technologies is rapidly and dynamically evolving; 2) students may have very diverse knowledge backgrounds, ranging from conventional information technology to data science, inducing various needs for inspiring skill and interaction engagement; and 3) while significant studies have been developed in understanding AI/ML-specific threats, most of the existing research focuses on computer vision domain with limited efforts made in the cybersecurity domain that is complex and rife with adversaries.

In (Sandhaus, Parreira and Ju, 2024)'s study of exploring students' self-initiated use of Generative AI tools in an interactive systems design class, the paper states that the impact of AI on software development is multifaceted. It affects coding practices, collaboration, and the role of data scientists, showing promise in accelerating programming.

Regarding the use of Generative AI in Human-Computer Interaction (HCI), students found it beneficial for tech-driven brainstorming, generating ideas for technology applications, and exploring new concepts and trends. They also used it as a prototyping assistant for coding and refining prototypes, enabling the generation of initial code snippets, and providing debugging solutions. Furthermore, Generative AI was efficiently utilized for documentation automation, which is crucial for any coding project. This is particularly important in cybersecurity too, where documenting and analyzing code, cleaning up text, expanding notes, and adding code comments are essential tasks.

In terms of synthesizing final feedback, Generative AI was used to analyze feedback and test results, allowing students to reflect on their own work and that of their teams. This capability is especially valuable when students collaborate on projects requiring coding and development, such as coursework.

The study of cyber security involves understanding different types of cyber-attacks, their mechanisms, and the challenges that need to be overcome. (Kumar et al., 2023). It primarily focuses on code analysis rather than code generation itself. Thus, Generative AI tools can help learners by providing code that they can analyse for any vulnerabilities. This aligns with its potential use in cybersecurity education, where these tools could assist individuals with no background knowledge by generating and understanding code for them.

Thus, it is crucial to consider the limitations of Generative AI for its safe integration in Cybersecurity Education. If the generated code is incorrect, the analysis will also be flawed, leading to incorrect language patterns and missed vulnerabilities. This could result in undetected security risks and potential malicious attacks.

(Michael, Abbas and Roussos, 2023)'s exploration of cybersecurity emphasises its various stages and states that the application of AI as a competitive advantage in organizational cybersecurity encompasses four key phases:

- Understanding the Cybersecurity Lifecyle: This phase involves comprehending the fundamental principles, tasks, and processes of the cybersecurity lifecycle. It forms the foundation of cybersecurity education, ensuring individuals understand the essential components and workflow of cybersecurity.

  Generative AI tools can facilitate this learning by recreating various cybersecurity scenarios, allowing students to engage interactively with the core components of the cybersecurity. Through AI-driven simulations and virtual environments, students can experience real-world cybersecurity challenges, helping them grasp foundational concepts more effectively.
- Data collection and aggregation from a variety of sources: The next phase focuses on gathering and integrating internal and external data from various sources which is crucial for a comprehensive understanding of the cybersecurity landscape, as it provides the base for analysis and threat detection.

  In this phase, Generative AI could be used to assist students generating various data scenarios, enabling students to practice aggregating and analysing information from multiple sources which is an essential part of identifying potential security threats.
- AI-enabled Analytics: After data collection, AI-enabled analytics are applied for identification of patterns and trends that may require urgent attention. AI's ability to process large volumes of data and make predictions ensures the recognition of potential cybersecurity threats.

Generative AI's capacity for advanced data analysis can be used by students to learn the application of these analytics. By using AI tools, students can learn to identify and predict cybersecurity threats, gaining hands-on experience with the same technologies used in the industry. This practical exposure is invaluable for understanding how to interpret and act on data insights.

- Knowledge Dissemination: The final phase involves using and sharing the knowledge gained to support cybersecurity goals. This ensures that the insights derived from AI analytics are effectively utilized to bolster security measures.

Generative AI can help students practice this by generating reports and documentation based on their analysis. AI can also expedite the dissemination of information in schools and universities, allowing students to understand the importance of clear communication and timely action in maintaining cybersecurity.

(Michael, Abbas and Roussos, 2023) further notes that this process aims to understand the data generated by assets and available metadata, enabling near real-time data analysis. This real-time analysis can provide timely insights into cybersecurity attacks, enhancing the ability to respond promptly.

These phases not only enhance industry practices but could also provide a robust structure for cybersecurity education, using Generative AI as a learning tool. It allows students to experience and engage with each phase interactively, from understanding basic principles to applying advanced analytics. This integration of generative AI into cybersecurity education ensures that students are not only aware of theoretical concepts but also proficient in practical applications, preparing them to tackle real-world cybersecurity challenges effectively.

(Kaur, Gabrijelčič and Klobučar, 2023) explore the use of AI in automated vulnerability identification and assessment, highlighting significant advancements in the field. Researchers have leveraged deep learning and transfer learning techniques to detect software vulnerabilities by analysing source code (Nembhard and Carvalho, 2019).These studies employ text-mining techniques to enhance machine learning-based vulnerability detection models, coupled with recommendation systems that assist programmers in writing secure code.

Additionally, Kaur, Gabrijelčič, and Klobučar (2023) discuss efforts to detect emerging vulnerabilities in software and cybernetic infrastructure through vulnerability repositories and social networks. A novel approach involves modelling the behavior of cyber-physical systems (CPS) and IoT devices under attack at both system and network levels, using machine learning to identify potential attack vectors (Saha et al., 2022).

AI-based fuzzing is another critical area where researchers are making strides. This technique involves injecting erroneous, unexpected, or randomly generated data into programs or interfaces and monitoring for crashes, failed code assertions, undocumented jumps, debug routines, and possible memory leaks. AI techniques help develop automated systems for identifying potential attack options, generating test cases, and analysing crashes. Researchers have utilized reasoning and natural language processing for seed generation techniques to increase code coverage with more unique execution paths, forming a fundamental part of smart fuzzing systems.

Test case generation is a widely studied application of AI-based fuzzing, with significant work done in web browsers, compilers, CPS, software libraries, and simple computer programs. Automated penetration testing is another area where AI shows promise, using reinforcement learning to exploit known or zero-day vulnerabilities in large networks and microgrid control algorithms. This intelligent approach aims to efficiently penetrate attack surfaces and assess what attackers can gain from current environments. Overall, the integration of AI in cybersecurity enhances the detection, assessment, and mitigation of vulnerabilities, making systems more robust against evolving threats.

5 Concerns of GenAI: Transparency, Accuracy and Security

The integration of GenAI tools to enhance the learning outcomes of Cybersecurity Education holds potential for students but there are certain challenges and limitations that should be addressed. This section explores the three concerns of GenAI, i.e., Transparency, Accuracy and Security.

5.1 Accuracy

Accuracy is essential within the GenAI models as inaccurate results could lead to misinformation among the students which is crucial, particularly in cybersecurity. Any false knowledge or incorrect code generation could lead to false interpretation and the inability to identify potential threats and vulnerabilities in the cybersecurity context.

One of the current research trends on AI and cybersecurity usually focuses on two different themes: the trustworthiness of AI systems and the application of AI in the cybersecurity domain. While it is wise for the researchers to have their own concentrations, it will be ill-advised to treat AI and cybersecurity as two distinct subjects to educate the next generation of scholars. (Wei-Kocsis et al., 2024) have proposed a virtual, proactive, and collaborative learning paradigm which aims to innovate cybersecurity and AI/ML curriculum by integrating two key components. First, it offers a cost-effective immersive learning environment that encourages students to explore AI/ML in real-world cybersecurity contexts by building models with tangible objects. Second, it introduces a game-based education approach through hackathons, enabling students from diverse backgrounds to collaboratively identify AI/ML-specific threats and develop trustworthy cybersecurity solutions.

A recent study (Zhao et al., 2023) has raised concerns about the accuracy of Generative AI (GAI) models like ChatGPT and Bard. The study discovered several issues with the factual claims made by these models:

- some claims contradict established reference sources,
- some claims cannot be traced back to any known sources, and
- others lack supporting references or conflict with information from multiple online sources. This leads to misinterpretation which can result in false learning among the students.

For example, the experiment resulted in Bing giving fabricated numbers in the summary of financial reports and Bard produced errors in its answer regarding which telescope captured the first exoplanet images.

(Zhao et al., 2023) highlight that AI models often produce conflicting or non-existent facts, even when provided with reliable sources. The generative nature of these models makes it difficult to control their outputs and ensure factual consistency. The research recommends imposing restrictions to prevent conversational AI from generating unsafe and inaccurate information.

In an experiment by Tlili et al. (2023), the generative AI tool ChatGPT was provided with two different papers by different authors, each offering distinct reviews on the broad topic of using chatbots in education. Despite the differences in the reviews and results of the papers, ChatGPT produced very generic and similar summaries for both, generating the same content regarding the benefits and challenges. This raises concerns about the reliability and accuracy of generative AI models for content analysis and their use in research and providing learning content in the field of education.

In their paper, (Osinga, 2005) discuss incorporating AI into the OODA (Observe, Orient, Decide, Act) methodology, which is used by machine learning specialists to implement commercial security lifecycles. This methodology comprises four stages:

- Observe: In this initial step, a variety of network traffic is monitored. Since human security analysts are not as capable of processing vast amounts of data traffic, machines make this observation and detection more efficient.
- Orient: The second step involves identifying emerging patterns that require some form of cybersecurity adaptation to combat potential threats.
- Decide: In the third step, AI in cybersecurity allows specialists to write algorithms that observe incoming and outgoing data traffic within a given context. This helps detect network-based traffic patterns and simultaneously report these findings through situational awareness databases.
- Act: The final step involves running the previously developed AI-based algorithms at short intervals without human involvement, but only when an action is required in response to a machine-determined alert linked to an exception.

By following this framework, AI algorithms can achieve 99% accuracy, with the remaining 1% managed by keeping humans involved in the loop. Despite the advantages and efficiency of integrating AI into the process, the strategic level perspective of the overall socio-technical system, within which the AI and cybersecurity strategies and methodologies are embedded, remains a critical consideration. This is particularly important in addressing security concerns that extend beyond the AI or technological elements alone (Michael, Abbas and Roussos, 2023).

5.2 Transparency

Another issue facing Generative AI models is the lack of transparency. Not only is there a risk of inaccuracy in the results produced by these models, but they also do not disclose the sources of their information. This creates a "black box" effect, preventing users from verifying the references that GAI models rely on. The lack of transparency when learning Cybersecurity could lead to the inability to understand all the steps behind the different concepts and code generation which is crucial when it comes to avoiding half information among students.

These models give the impression that the AI itself is providing answers, while the sources remain hidden. (Olga *et al.*, 2023). They also invent non-existent facts when needed to complete a plausible text (Munn, Magee and Arora, 2023). It may happen that some or even most of a response to a prompt is fact, but the machine has no way of distinguishing fact from fake in its sources. These are genre machines, harvesting ostensible facts they have found in their textual sources but without being able to verify them (Olga *et al.*, 2023).

In a study by Tlili et al. (2023), participants were interviewed about the use of generative AI in education. Several ethical concerns were raised, including issues of plagiarism, random inaccuracies, and the vagueness of ChatGPT on relevant topics. Participants noted that the output generated by the tool often seemed more like opinion without references rather than factual information. This highlighted significant concerns regarding the transparency of generative AI tools, as they can produce incorrect information without revealing their sources. This lack of transparency can lead to misinformation and trust issues among students if they use these tools in education.

(Adadi and Berrada, 2018) suggest that for Generative AI models to develop into trustworthy platforms, they must not remain inscrutable black boxes. These models should be fully transparent about their data sources and potential biases. Additionally, they should report when they are low in confidence in their answers and explain the reasons behind this uncertainty.

(Lin, Hilton and Evans, 2022) demonstrate how GPT-3 can express uncertainty and confidence about its answers in natural language without relying on model logits. The motivation behind this calibration is to address the challenge of truthfulness in language models.

(Mukhoti et al., 2020) argue that if generative AI models convey calibrated uncertainty about their statements, it enhances transparency by allowing users to understand the confidence level of each response. This transparency means that users can assess the reliability of the information based on the model's expressed certainty. By knowing how much trust to place in a given statement, users can make more informed decisions and use the AI more effectively. This approach openly communicates the model's confidence, fostering trust and better interaction between users and AI systems.

This transparency is particularly important for current models, which often hallucinate falsehoods, and for any model that makes statements where there is no known ground truth, such as economic forecasts or open problems in science or mathematics. By indicating their uncertainty, models can help users navigate these complex and uncertain domains with greater confidence and accuracy.

5.3 Security

Security is vital when it comes to knowledge dissemination especially in Cybersecurity education because of the sensitive data in this field. If the GenAI models themselves aren't secure, learning from these tools would lead to issues of safeguarding student's personal information and online safety.

(Wei-Kocsis et al., 2024) presents a methodology for leveraging AI in code review and code generation with specific focus on security bugs detection as manual code review could be an extensive process, vulnerable to human errors. It's suggested that by training

GPT-4 with a vast dataset of past code reviews and known security vulnerabilities across different languages, it can act as an automated code reviewer, capable of identifying potential security bugs across various programming languages.

(Wei-Kocsis et al., 2024) discusses how these Generative AI models can be attacked to bypass its ethical and privacy safeguards using reverse psychology and jailbreak techniques and ChatGPT's defence mechanisms to support it. But there is still some research to be done on the validity and thoroughness of these mechanisms. In terms of education, the paper suggests that specifically, ChatGPT can be trained to understand better and respond to each user's specific needs and preferences. Furthermore, by leveraging the vast amounts of data generated by ChatGPT's interactions, developers can create language models that are highly tuned to each user's specific needs and preferences, leading to a more personalized and engaging experience.

In the study conducted by Tlili et al. (2023), privacy was identified as a major concern. The paper discusses the experiment's findings, highlighting certain claims made by the official OpenAI website:

- "We may use your content to train our models."
- "We review conversations to improve our systems and to ensure the content complies with our policies and safety requirements."

Although OpenAI offers options to opt out of using conversations for training and reinforces its Privacy Policy and Terms of Use, it is crucial to note that users, including young people and those from technical backgrounds, might develop a level of trust and emotional attachment to the tool (Tlili et al., 2023). This trust may lead them to overlook privacy concerns and feel comfortable sharing personal information.

Furthermore, the website lacks transparency about how it reviews and uses this information. When asked about this, ChatGPT contradicted the FAQ on the website and refused to explain how the information is used and stored. This raises significant concerns about privacy and the cyber footprint left by students using generative AI models regularly (Pencheva, Hallett and Rashid, 2020).

In (Law, 2024)'s scoping review (Zhu et al., 2023)'s emphasis on the ethical considerations concerning data privacy and security is mentioned. They discuss the risks associated with security breaches and hacking, which could result in unauthorized access to students' data and its exploitation by third parties. The importance of safeguarding personal information through obtaining prior consent before utilizing Generative AI tools is highlighted.

Furthermore, (Law, 2024) discusses the lack of AI literacy and awareness among teachers, highlighting that this could lead them to unknowingly introduce advantageous Generative AI tools to students without fully understanding how first-party data, rather than third-party data, is collected and utilized. The issue of insufficient training for teachers in cybersecurity is also addressed by (Pencheva, Hallett and Rashid, 2020). Both sources mention the tension between teachers' willingness to teach cybersecurity and the lack of subject-specific, ready-made resources available for this purpose.

In an experiment conducted by Pencheva, Hallett, and Rashid (2020), participants highlighted that schools often do not distinguish between personal safety and cybersecurity and may not be adequately equipped to support students in facing these challenges. The paper emphasizes that both secondary and primary school teachers frequently lack relevant knowledge and are uncertain about how to engage or teach even basic cybersecurity concepts.

This reveals a significant gap: although teachers are enthusiastic and willing to teach cybersecurity, they lack the necessary support and knowledge to effectively convey this information to students. As a result, students are not informed about potential online threats or the importance of cybersecurity in protecting themselves and exploring future career opportunities in this field.

(Zhao et al., 2023) highlight that, in addition to generating inaccurate outputs, generative AI models can produce unsafe content. Malicious parties can eventually bypass safety restrictions, and fact verification remains an unresolved challenge.

Thus, addressing these themes is important for the successful implementation of GenAI in Cybersecurity Education, ensuring that these tools enhance learning experience.

6 Existing Work Addressing the Limitations of Generative AI

Some existing work has been done to make sure these GenAI models produce accurate and transparent results. Research from University of Oxford (Lin, Hilton and Evans, 2022) demonstrate how GPT-3 can express uncertainty and confidence about its answers in natural language without relying on model logits. Upon receiving a prompt, the chatbot produces an answer along with a confidence level, such as 90% or "high confidence." These confidence levels are well-calibrated probabilities reflecting the tool's own answers. The motivation behind this calibration is to address the challenge of truthfulness in language models. In this experiment, three types of probabilities were considered: Verbalized, Answer Logit, and Indirect Logit. According to the paper, as Answer Logit and Indirect Logit are log-based probabilities, they cannot be used for models without traceable outputs. However, verbalized probabilities, which apply to models that output natural language and mirror human expressions of certainty, are particularly useful. This approach is beneficial for non-technical users or students learning a new topic, as it provides the transparency they expect.

One of the other methods was InstructGPT which aligns language models with user intent on a wide range of tasks by fine-tuning with human feedback. Starting with a set of labeller-written prompts and prompts submitted through the OpenAI API, (Ouyang et al., 2022)collected a dataset of labeller demonstrations of the desired model behaviour, which was used to fine-tune GPT-3 using supervised learning. A dataset of rankings of model outputs was then collected, which was used to further fine-tune the supervised model using reinforcement learning from human feedback. (Ouyang et al., 2022). In this procedure, a team of forty contractors were first hired, selected based on their performance in a screening test. These contractors label the data. Next, a dataset of human-written demonstrations of the desired output behaviour was gathered in response to prompts submitted to the OpenAI API, along with some prompts written by the labellers. This dataset was then used to train the supervised learning baselines. (Kasneci et al., 2023).

Following this, a dataset of human-labelled comparisons between outputs from the models on a larger set of API prompts was collected. This dataset was used to train a reward model (RM) that predicts which model output the labellers would prefer. Finally, the RM was employed as a reward function to fine-tune the supervised learning baseline to maximize the reward using the PPO algorithm.

(Olga *et al.*, 2023) states that rather than criticizing specific models or companies, they hope to call on researchers and developers to focus on improving the transparency and factual correctness of AI services, allowing humans to place a higher level of trust in the foreseeable future. But is this the only possible solution? Still having human factor in these GAI models would contradict the purpose of having these models in the first place. Thus, there is a need of a framework that is accurate, transparent, secure and validates the results at the same time.

7 Conclusion

There is an education and training gap to foster a qualified cyber-workforce that understands the usefulness, limitations, and best practices of AI technologies in the Cybersecurity domain. How can educators and students navigate the challenges posed by the lack of transparency in Generative AI models? Moreover, what measures can be implemented to enhance the security and ethical use of Generative AI in educational settings?

By examining these questions and exploring solutions, this paper aimed to shed light on the potential of Generative AI in Cybersecurity education while identifying key research gaps and avenues for future exploration. The integration of GenAI in Cybersecurity education indicates promising enhancement of dissemination of cybersecurity concepts to students but also introduces challenges in the themes of transparency, accuracy, and security. Addressing these challenges becomes crucial while exploring this integration so that the ethical standards and educational integrity are not compromised.

Thus, as discussed, while the genie is out of the bottle, it is important to guide its magic safely and securely and focus on the question once again: Can Generative AI be safely incorporated as an assistive tool for Cybersecurity education?

Future work focuses on researching frameworks that ensure the safe assimilation of GenAI in Cybersecurity education so that it could be used in a transparent, accurate and secure way. Examining how GenAI can be used in educational settings, considering the existing limitations of the tool and the gaps in the Cybersecurity education domain would lead to a reliable and trustworthy use of AI models, making them more effective as educational tools.

By continuing to explore these areas, students could be introduced to new and existing Cybersecurity concepts so that they are aware of the cyber threats and vulnerabilities and contribute to a more secure digital environment.

References

Ghaleb, A.F., et al.: Misbehavior-aware on-demand collaborative intrusion detection system using distributed ensemble learning for VANET. Electronics **9**(9), 1411 (2020). https://doi.org/10.3390/electronics9091411

Adadi, A., Berrada, M.: Peeking inside the black-box: a survey on explainable artificial intelligence (XAI). IEEE Access **6**, 52138–52160 (2018). https://doi.org/10.1109/ACCESS.2018.2870052

Artificial Intelligence (AI) in education conference (2024) Intranet - Cardiff University. https://intranet.cardiff.ac.uk/staff/news/view/2818378-artificial-intelligence-ai-in-education-mini-conference. Accessed 21 Jul 2024

Baidoo-Anu, D., Ansah, L.O.: (no date) Education in the Era of Generative Artificial Intelligence (AI): Understanding the Potential Benefits of ChatGPT in Promoting Teaching and Learning

Brown, T., et al.: Language models are few-shot learners. In: Advances in Neural Information Processing Systems. Curran Associates, Inc., pp. 1877–1901 (2020) . https://proceedings.neurips.cc/paper/2020/hash/1457c0d6bfcb4967418bfb8ac142f64a-Abstract.html. Accessed 17 Jul 2024

Catal, C., et al.: Analysis of cyber security knowledge gaps based on cyber security body of knowledge. Educ. Inf. Technol. **28**(2), 1809–1831 (2023). https://doi.org/10.1007/s10639-022-11261-8

ChatGPT (no date). https://openai.com/chatgpt/. Accessed 22 Jul 2024

Children and parents: media use and attitudes report 2022 (no date)

Chiu, T.K.F., et al.: Systematic literature review on opportunities, challenges, and future research recommendations of artificial intelligence in education. Comput. Educ. Artif. Intell. **4**, 100118 (2023). https://doi.org/10.1016/j.caeai.2022.100118

Grover, S., Broll, B., Babb, D.: Cybersecurity education in the age of AI: integrating AI learning into cybersecurity high school curricula. In: Proceedings of the 54th ACM Technical Symposium on Computer Science Education V. 1. SIGCSE 2023: The 54th ACM Technical Symposium on Computer Science Education, Toronto ON Canada: ACM, pp. 980–986 (2023). https://doi.org/10.1145/3545945.3569750

Haleem, A., Javaid, M., Singh, R.P.: An era of ChatGPT as a significant futuristic support tool: a study on features, abilities, and challenges. BenchCouncil Trans. Benchmarks Stand. Evaluations **2**(4), 100089 (2022). https://doi.org/10.1016/j.tbench.2023.100089

Hsu, Y.-C., Ching, Y.-H.: Generative artificial intelligence in education, part one: the dynamic frontier. TechTrends **67**(4), 603–607 (2023). https://doi.org/10.1007/s11528-023-00863-9

Kasneci, E., et al.: ChatGPT for good? On opportunities and challenges of large language models for education. Learning and Individual Differences (2023) [Preprint]

Kaur, R., Gabrijelčič, D., Klobučar, T.: Artificial intelligence for cybersecurity: literature review and future research directions. Inf. Fusion, **97**, 101804 (2023). https://doi.org/10.1016/j.inffus.2023.101804

Kumar, G., et al.: Cybersecurity education: understanding the knowledge gaps based on cyber security policy, challenge, and knowledge. In: 2023 IEEE 12th International Conference on Communication Systems and Network Technologies (CSNT). 2023 IEEE 12th International Conference on Communication Systems and Network Technologies (CSNT), pp. 735–741 (2023). https://doi.org/10.1109/CSNT57126.2023.10134610

Law, L.: Application of generative artificial intelligence (GenAI) in language teaching and learning: a scoping literature review. Comput. Educ. Open **6**, 100174 (2024). https://doi.org/10.1016/j.caeo.2024.100174

Lin, S., Hilton, J., Evans, O.: Teaching Models to Express Their Uncertainty in Words (2022). arXiv. http://arxiv.org/abs/2205.14334. Accessed 20 May 2024)

McIntosh, T., et al.: Harnessing GPT-4 for generation of cybersecurity GRC policies: a focus on ransomware attack mitigation. Comput. Secur. **134**, 103424 (2023). https://doi.org/10.1016/j.cose.2023.103424

Meli, K., Taouki, I., Pantazatos, D.: EMPOWERING EDUCATORS WITH GENERATIVE AI: THE GENAI EDUCATION FRONTIER INITIATIVE (2024)

Michael, K., Abbas, R., Roussos, G.: AI in cybersecurity: the paradox. IEEE Trans. Technol. Soc. **4**(2), 104–109 (2023). https://doi.org/10.1109/TTS.2023.3280109

Mukhoti, J., et al.: Calibrating Deep Neural Networks using Focal Loss. In: Advances in Neural Information Processing Systems. Curran Associates, Inc., pp. 15288–15299 (2020). https://proceedings.neurips.cc/paper/2020/hash/aeb7b30ef1d024a76f21a1d40e30c302-Abstract.html. Accessed 13 Jul 2024

Munn, L., Magee, L., Arora, V.: Truth Machines: Synthesizing Veracity in AI Language Models (2023). arXiv. http://arxiv.org/abs/2301.12066. Accessed 26 Apr 2024)

Munn, Z., et al.: Systematic review or scoping review? Guidance for authors when choosing between a systematic or scoping review approach. BMC Med. Res. Methodol. **18**(1), 143 (2018). https://doi.org/10.1186/s12874-018-0611-x

Nembhard, F., Carvalho, M.: The impact of interface design on the usability of code analyzers. In: 2019 SoutheastCon. 2019 SoutheastCon, pp. 1–6 (2019). https://doi.org/10.1109/SoutheastCon42311.2019.9020339

Nowrozy, R., Jam, D.: Embracing the Generative AI Revolution: Advancing Tertiary Education in Cybersecurity with GPT (2024). arXiv. http://arxiv.org/abs/2403.11402. Accessed 29 May 2024

Olga, A., et al.: Generative AI: implications and Applications for Education (2023). https://doi.org/10.48550/ARXIV.2305.07605

Online FlowChart & Diagrams Editor - Mermaid Live Editor (no date). https://mermaid.live. Accessed 22 Jul 2024

Osinga, F.: Science, strategy and war: the strategic theory of John Boyd. Eburon Academic Publishers, Delft (2005)

Ouyang, L., et al.: Training language models to follow instructions with human feedback (2022). arXiv. Available at: http://arxiv.org/abs/2203.02155. Accessed 25 Apr 2024

Pencheva, D., Hallett, J., Rashid, A.: Bringing cyber to school: integrating cybersecurity into secondary school education. IEEE Secur. Priv. **18**(2), 68–74 (2020). https://doi.org/10.1109/MSEC.2020.2969409

Quayyum, F., Cruzes, D.S., Jaccheri, L.: Cybersecurity awareness for children: a systematic literature review. Int. J. Child Comput. Interact. **30**, 100343 (2021). https://doi.org/10.1016/j.ijcci.2021.100343

Sag, R.B.: A systematic literature review on cyber security education for children. IEEE Trans. Educ. **66**(3), 274–286 (2023)

Sahu, N.: (no date) THE GENAI REVOLUTION: UNLEASHING THE ROLE OF INFORMATION TECHNOLOGY IN EDUCATION (2583)

Sandhaus, H., Parreira, M.T., Ju, W.: Student Reflections on Self-Initiated GenAI Use in HCI Education (2024). arXiv. http://arxiv.org/abs/2405.01467. Accessed 12 Jul 2024

Student perceptions of generative AI report (no date)

Thurzo, A., et al.: Impact of artificial intelligence on dental education: a review and guide for curriculum update. Educ. Sci. **13**(2), 150 (2023). https://doi.org/10.3390/educsci13020150

Tlili, A., et al.: What if the devil is my guardian angel: ChatGPT as a case study of using ChatBots in education. Smart Learn. Environ. **10**(1), 15 (2023). https://doi.org/10.1186/s40561-023-00237-x

Tricco, A.C., et al.: PRISMA extension for scoping reviews (PRISMA-ScR): checklist and explanation. Ann. Internal Med. **169**(7), 467–473 (2018). https://doi.org/10.7326/M18-0850

Trust, T., Whalen, J., Mouza, C.: (no date) Editorial: ChatGPT: Challenges, Opportunities, and Implications for Teacher Education

Wei-Kocsis, J., et al.: Cybersecurity education in the age of artificial intelligence: a novel proactive and collaborative learning paradigm. IEEE Transactions on Education, pp. 1–0 (2024). https://doi.org/10.1109/TE.2023.3337337

Wohlin, C.: Guidelines for snowballing in systematic literature studies and a replication in software engineering. In: Proceedings of the 18th International Conference on Evaluation and Assessment in Software Engineering. EASE '14: 18th International Conference on Evaluation and Assessment in Software Engineering, London England United Kingdom: ACM, pp. 1–10 (2014). https://doi.org/10.1145/2601248.2601268

Wohlin, C., et al.: Successful combination of database search and snowballing for identification of primary studies in systematic literature studies. Inf. Softw. Technol. **147**, 106908 (2022). https://doi.org/10.1016/j.infsof.2022.106908

Xu, W., Ouyang, F.: The application of AI technologies in STEM education: a systematic review from 2011 to 2021. Int. J. STEM Educ. **9**(1), 59 (2022). https://doi.org/10.1186/s40594-022-00377-5

Zhao, R., et al.: Can ChatGPT-like Generative Models Guarantee Factual Accuracy? On the Mistakes of New Generation Search Engines (2023). arXiv. http://arxiv.org/abs/2304.11076. Accessed 26 Feb 2024

Zhu, I.C., et al.: How to harness the potential of ChatGPT in education?, vol. 15, pp. 133–152 (2023). https://doi.org/10.34105/j.kmel.2023.15.008

How I Created 63 Digital Forensics Cases: Experiences in the Concurrent Generation of Multiple Digital Forensic Scenarios

Harjinder Lallie(✉)

University of Warwick, Coventry, UK
hl@warwick.ac.uk

Abstract. Academic tutors aim to provide students with realistic teaching and assessment experiences in digital forensic investigation, where digital forensic images (DFIs) are crucial. DFIs are essential for understanding operating systems, file structures, and application behaviours, serving both educational and research purposes. While DFIs can teach various skills, this chapter focuses on those specifically used for training in incident investigation, whether corporate or criminal, referred to as *investigative DFIs*. Creating *investigative DFIs* is complex, requiring researchers to ensure contents are realistic and free from real data that could breach data protection laws or pose legal risks. Despite the availability of many DFIs, there is a scarcity of publicly accessible *investigative DFIs* suitable for assessment tasks. Available DFIs often have their solutions widely disseminated online, reducing their educational value. This situation compels academics to develop their own DFIs, a demanding task requiring the creation of credible, realistic scenarios that align with investigative learning objectives. The development process is time-consuming and requires meticulous attention to detail to maintain a plausible narrative. This chapter explores the challenges of developing effective *investigative DFIs*, sharing insights from the authors' experiences and proposing a six-step framework for developing multiple *investigative DFIs* concurrently. A case study illustrates the application of this framework.

1 Introduction

Academic tutors aim to provide realistic teaching and assessment experiences of digital forensic investigation. Digital forensics images (DFI) – bit for bit copies of a computer or mobile system such as a laptop, mobile device etc. – are integral to teaching and assessment. DFIs help learners understand operating systems, file systems, and the characteristics of applications, and are also invaluable for researchers who can test theories and tools accordingly [Lallie et al. (2011) Lallie, Lawson, and Day].

Although DFIs can be used to teach multiple skills, the focus of this chapter is on DFIs which can be used to teach learners how to conduct an investigation

P. Legg et al. (Eds.): CSE 2024, LNNS 1213, pp. 22–38, 2024.
https://doi.org/10.1007/978-3-031-77524-6_2

into an incident - corporate or criminal. We refer to these DFIs as *investigative DFIs*, i.e. DFIs used to teach investigation skills.

Developing *investigative DFIs* is not a trivial task. Researchers have to ensure the contents of the DFI are not only realistic, but that the DFI does not contain any 'real data', personal identities, or other data that could create data protection or other legal challenges. Consequently, although there are innumerable published DFIs (described in Sect. 2.2), there is a paucity of publicly available DFIs that could be used to set investigative assessment tasks, and where *investigative DFIs* are available publicly, the answers are either already out in the wild – published on websites and even videos, or will be published soon after the DFI is released [ForensicXS (2020)].

This leaves little option but for academics to self-develop DFIs. However, this is by no means an easy task because *Investigative DFIs* must be realistic, and the case scenarios contained therein to be credible. They have to be developed very carefully so as to maintain a plausible narrative. Consequently, they take a long time to develop.

This chapter examines the problem of creating effective *investigative DFIs*. We share experiences of developing DFIs, and propose a six step framework for developing multiple *investigative DFIs* concurrently. We developed multiple *investigative DFIs* so they could be used across several modules for both first sit and resit assessment components. Each assessment required a unique case.

We also present a case study which demonstrates the application of the framework.

2 Background

We begin by reviewing previous research into the use of DFIs. The review will show that, despite the importance of DFIs in academia, there has been little focus applied to the development of *investigative DFIs*. We also describe some of the DFI collections available for academic research use.

2.1 Literature Survey

The paucity of DFIs was first highlighted by [Lallie (2010)] who explored the development and use of DFIs. The study categorised and examined various DFIs, and emphasised the lack of DFIs suitable for assessment purposes. [Lallie (2010)] outlined the importance of inserting evidentiary and non-evidentiary artefacts within the DFI – the latter referred to as 'noise' and discussed further in Sect. 3.2.2. The study emphasised the difficulty of creating DFIs and proposed DFI creation strategies highlighting the use of personally generated vs learner generated DFIs.

[Horsman and Lyle (2021)] underlined the shortage of suitable DFIs and emphasised that the key challenge is the acquisition of realistic data that accurately represents various digital environments and user behaviours without violating privacy laws or ethical standards. The cases need to be authentic whilst ensuring anonymisation - both very difficult tasks.

[Horsman and Lyle (2021)] stress the importance of standardised methodologies and collaboration between stakeholders, including researchers, law enforcement, and industry experts, to overcome these obstacles. [Horsman and Lyle (2021)] conclude by advocating for improved frameworks and best practices in dataset construction, which are crucial for advancing the field of Digital Forensics and ensuring more robust and reliable forensic analyses.

[Grajeda et al. (2017) Grajeda, Breitinger, and Baggili] underscored significant deficiencies which include:

- The lack of realistic and large-scale datasets;
- A limited representation of various operating systems and file types; and
- The scarcity of datasets that reflect contemporary cyber threats.

The solution to the problem, argue [Grajeda et al. (2017) Grajeda, Breitinger, and Baggili] is a collaborative effort between academia, industry, and government agencies to create and share more comprehensive and realistic datasets. Moreover, [Grajeda et al. (2017) Grajeda, Breitinger, and Baggili] argue the need for a standardised protocol for dataset creation.

[Breitinger and Jotterand (2023)] argue that there are complex legal implications in sharing datasets within the Digital Forensics community. The authors proceed to propose a taxonomy to categorise different types of digital forensic datasets used. The taxonomy categorises datasets based on their source, purpose, and sensitivity.

[Breitinger and Jotterand (2023)] relate to privacy, data protection, and intellectual property rights. There is a risk, the authors argue, of exposing sensitive personal information and violating legal regulations.

[Breitinger and Jotterand (2023)] advocate for increased collaboration between legal experts and practitioners to develop guidelines that ensure datasets can be shared without compromising legal and ethical standards.

2.2 Availability of DFIs

Numerous DFIs have been published online by entities such as: Computer Forensic Reference Data Sets - CFReDS [NIST (2023)], Digital Corpora [Garfinkel and Dinolt (2009)], and Digital Forensics Tool Testing Images [Carrier (2010)]. In addition to these, there are numerous DFIs available in private collections created by tool designers, academics, and researchers.

The CFReDS is one of the largest publicly available datasets containing 236 unique DFIs at the time of writing, dating from 1950 to the present day. The dataset has been gathered through various sources which include DFRWS (Digital Forensics Workshop) challenges; various universities including: Columbia, Florence, and Göttingen; and companies such as Belkasoft and Cellebrite.

The Digital Forensics Tool Testing Images published by [Carrier (2010)] were developed specifically for that purpose - aiding tool testing. This dataset contains DFIs that can be used to test extended partitions, FAT systems (keyword searching), NTFS systems, EXT3FS systems, JPEG searching, and basic data carving.

We can see from this, that there are generally three types of DFI, they can be categorised as:

- *Investigative DFIs* are designed to assess a learner's ability to conduct a realistic incident/criminal investigation. These incorporate pseudo-corporate/criminal incidents placed within the context of a fictional scenario.
- *Skill specific DFIs* are designed to assess specific technical skills in a learner. For example, the DFI might test the ability to carve files, deconstruct a file system, investigate a registry.
- *Tool testing DFIs* are used to test tool functionality. Often the tool has a specific purpose, and the DFI has been built with pre-planted data aimed at testing that tool.

2.3 Summary

This brief review reveals that although there are many DFIs available in the public domain, very few are suitable for use in an assessment scenario because the answers are likely to be available online.

There exists a lack of frameworks to create DFIs [Lallie(2010), Grajeda et al.(2017)Grajeda, Breitinger, and Baggili, Horsman and Lyle(2021)]. These frameworks can aid researchers to create new cases. One of the difficulties in creating a DFI relates to the ethical and legal challenges [Horsman and Lyle (2021), Breitinger and Jotterand (2023)]. Added to this are the lack of different operating systems/file types, and the scarcity of datasets that reflect contemporary cyber threats.

The present study focusses proposes a framework for creating a DFI. Researchers can take the guidance provided herein to create their own DFIs

3 Design

This section describes a framework that can aid the self-development of *investigative DFIs*. We describe each sage of the framework and highlight important components relating to the development of *investigative DFIs* such as the importance of: meticulous record keeping, the insertion of noise, and the insertion of trip hazards.

3.1 The Framework

The framework comprises of six stages: *narrative development, email creation, OS install, app install and configuration, enactment* and *DFI and case statement creation*. These are described below and illustrated in Fig. 1.

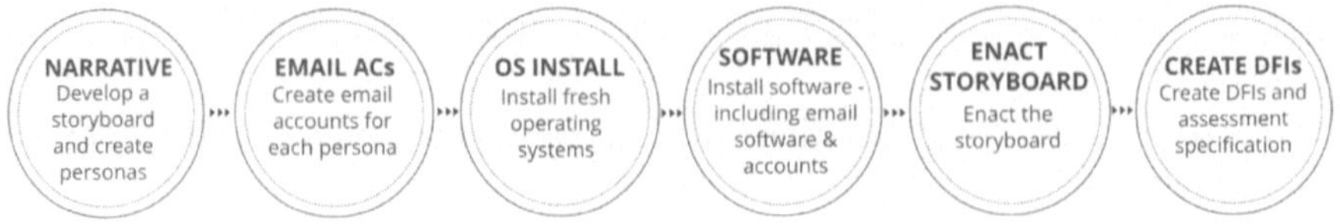

Fig. 1. The Investigative DFI Creation Framework

3.1.1 Stage 1, Narrative Development The *narrative* is a high level overview of the incident(s) to be investigated. It is a story-line which includes the key characters, their personas, and the incident. Narrative development requires good creativity and story writing skills. Stories have to be credible, engaging, but not dramatic. That said, a small credible twist inspires creativity and keeps learners engaged. In our example, the twist was an assault on a drug distributor.

The narrative describes the *incident or incidents* to be investigated, the *user accounts* to be created, the *personas* involved in the incident(s), and the *noise* that needs to be generated within the DFI.

The *incident* or *incidents* are the issue under investigation. There can be more than one incident described in parallel. Three sample incidents – the possession, distribution, and creation of illegal images, drug trafficking, and an assault – are described in Sect. 4.

A *persona* is a fictional person. A persona has a job, hobby, interests, etc. we refer to this as their 'profile'. The profile might include personal characteristics and features, for example, two personas on an operating system might be husband and wife. Their profile might include activity such as booking a holiday, looking for a new home.

Personas will fall into a few categories, those that have:

- A user account on a computer, and an email account.
- An email account, but no user account – or at least if they do have a computer, it is not available to us (because we did not create it).
- Neither a user account or an email account. They are mentioned in communications.

It is useful to include an 'innocent persona' within this as this tests the investigator's ability to apply more scrutiny to the case, particularly if the 'innocent persona' is set as the target of the investigation. When the assessment is provided to the learner, they will not of course be made aware that they are investigating an innocent persona.

For an investigation to incorporate some sense of reality, there needs to be an adequate level of peripheral activity not linked to the incident(s) under investigation. We refer to this as 'noise' and discuss it in Sect. 3.2.2.

A key problem with developing the narrative relates to when we should stop, when do we have enough of a narrative to warrant sufficient investigatory challenge?

In our experience, thirty days of activity is adequate. More than this can lead to researcher fatigue and increases the likelihood of errors in the narrative.

In a case involving illegal images, the presence of 2,000 images is adequate and there is little benefit in adding more. In a case largely reliant on email communication, a total of 100 emails is sufficient, particularly if there are multiple personas and other peripheral activity (downloading of maps, distribution of maps/directions).

A sample of how we developed the narrative is provided in Fig. 2. We used sequence diagrams with each persona set out in the diagram. Our narrative did not specify exact communications, these were presented at a high level with the exact content left to be completed during the narrative enactment stage, The narrative was developed in a manner that it could be given to a third party to enact.

3.1.2 Stage 2, Email Account Creation We recommend creating email accounts away from the target devices. This makes the case more realistic as it would appear strange for multiple accounts to be created simultaneously on the target systems. This also ensures that any problems in creating email accounts are dealt with away from the target device.

We purchased a domain which enabled us to create a near limitless number of email addresses at the same time ensuring that we did not contravene any terms/conditions of email suppliers such as Gmail and Hotmail/Microsoft.

3.1.3 Stage 3, OS Install The operating system should be installed within a virtual machine with snapshots taken frequently to enable a roll back if there is an error.

We configured five operating systems. Once the OS was installed, we created accounts for each persona connected with a computer system. This is described in Fig. 3 where the dotted line represents computers and icons within those lines represent the accounts that were created.

3.1.4 Stage 4, Program/app Install Specific apps now need to be installed and configured. These include browsers and email clients. The choice of email client is important. We selected Thunderbird rather than using web-mail. Thunderbird and other desktop based clients keep evidence on the computer rather than where a web mail client is being used. We configured this with POP settings so that emails were downloaded to the client and stored locally.

We also installed open source photo editors to view and edit photos.

3.1.5 Stage 5, Enact Narrative Once the environment is established, we can begin enacting the narrative. This can be quite challenging as the communications, browsing and app activity needs to be enacted in a credible manner. Sequencing and consistent communication tone and style are important.

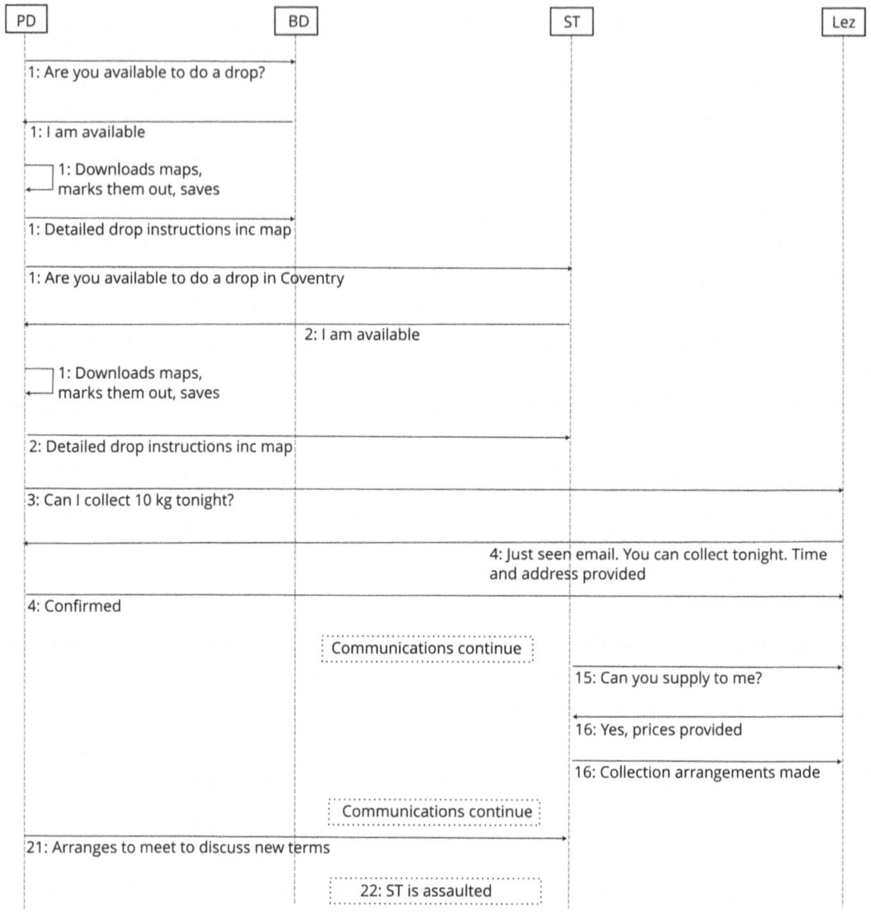

Fig. 2. Narrative development using sequence diagrams. Numbers correspond to day numbers. e.g. 15 is day 15

Enacting the narrative requires logging into user accounts, sometimes on several different VMs at the same time, and enacting the activity described in the narrative. A well constructed narrative can make this as easy as following the steps within the narrative.

3.1.6 Stage 6, DFI and Case Statement Creation Once the narrative has been enacted, a DFI can be taken of each computer system involved. So, for the example provided in Fig. 3, five DFIs were created.

3.2 Other Factors

3.2.1 Record Keeping A detailed record of activity in each user account must be kept at the same time as enacting the narrative. This is so that a marking scheme can be set. The record complements and extends the narrative. e.g. the narrative might say *'send an email which tells persona 2 to...'*, the record keeping will copy paste the contents of the email and record the time it was sent.

Other details that should be recorded at this point are the:

- Operating system, including the build number.
- Date and time of operating system installation and installation details of other software.
- Timezone.
- Specific login-logout, turn on/turn off times.
- Details of all accounts created on the system along with passwords. The passwords are not for learners, but the researcher.
- Email accounts created along with passwords. Passwords are not for the learner, but the researcher.
- Internet activity including: websites visited, bookmarked, files downloaded.

3.2.2 Noise A standard operating system installation with four to eight weeks of normal use will generate a significant level of non-evidentiary artefacts which we refer to as 'noise' [Lallie (2010)].

Without noise, the investigation may appear superficial and over-simplistic making evidentiary elements easily distinguishable, so noise is essential to the assessment as it adds a sense of realism to the investigation. Moreover, the addition of noise adds complexity and increases the investigation scope, e.g. subscriptions to newsletters will introduce text, URLs, and images into the investigation, whilst these are redundant to the investigation, they have to be investigated nevertheless.

It may not be possible to devote one's time to the computer system for four to eight weeks just to develop the 'noise', hence, noise has to be generated artificially through: the downloading/copying of photos, 'typical' web searches relating to the persona's areas of interest, and subscriptions to newsletters and blogs - which will ensure a trickle of incoming emails, associated text, and images. In any activity, for example subscriptions to newsletters, one must ensure the terms/conditions of the subscription are not contravened, e.g. the researcher may not be able to use a fake email address.

3.2.3 Trip Hazards We have found it useful to insert deliberate trip hazards within the DFI so as to test the diligence of the investigator. Trip hazards can be simple, for example subtly misspelt directory/file/persona names.

The researcher might deliberately set a timezone that does not reflect the timezone at which a persona was arrested. For example, although the timezone at which the suspect was arrested was: GMT Standard Time (UTC), the researcher might want to set the timezone on the computer to be: Eastern Standard Time (UTC-05:00). Assuming the learner even notices this discrepancy, this will introduce a small challenge in creating an accurate timeline.

Perhaps the ultimate trip hazard however, is the insertion of an innocent persona amongst the 'guilty' personas, and setting the innocent persona as the subject of investigation. Although an excellent test of the learner's ability to apply rigour and reasoning within an investigation, our experience has shown that this presents three challenges, learners might: become aggravated by not being able to find the evidence, they might argue that there is a mistake in the coursework specification 'the coursework is faulty', and/or they might incorrectly advocate guilt.

4 A Case Study

We now describe a sample case study to demonstrate how the framework described in Sect. 3 can be applied. Throughout the discussion the reader should refer to Fig. 3 which describes the computer systems (dotted lines) and user accounts created in our case study, and Fig. 4 which provides a social network of connections within the case study.

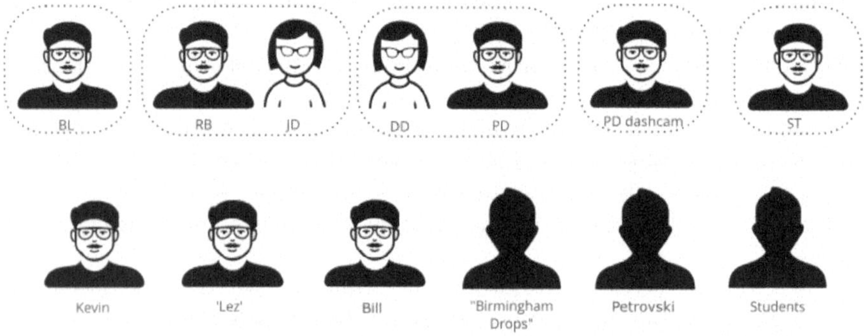

Fig. 3. DFIs created for the case study

We developed three concurrent fictional scenarios relating to the possession, distribution, and creation of illegal images; drug trafficking; and an assault.

Personas were created and linked with each case. User accounts were created for some personas, other personas were named in communications, but did not have a corresponding user account. These are described in Fig. 3 wherein personas with a dotted line around them are on the same computer system and have an associated DFI. Personas without a dotted line around them are named in communications or have sent emails to other personas, however, there is no associated DFI.

4.1 The Illegal Images Case

4.1.1 Illegal Images: Personas Three personas were associated with the **illegal images** case: Denise Djanko (DD), Rueben Beethomahler (RB), and

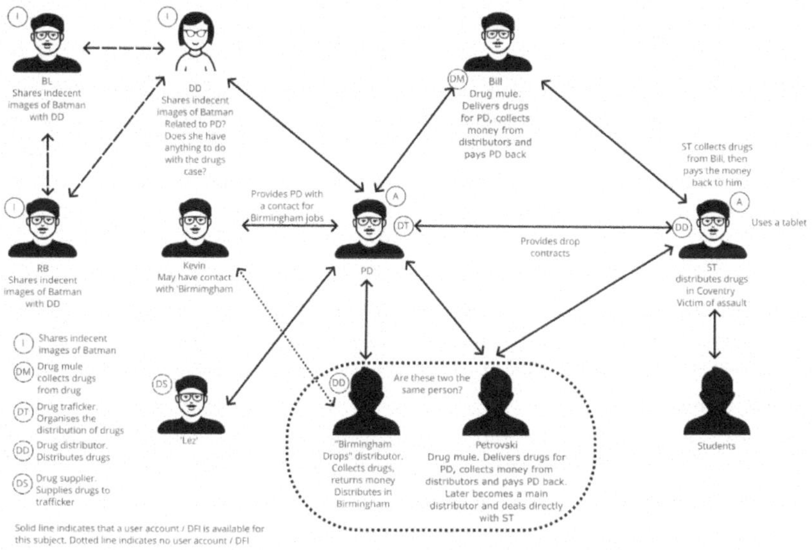

Fig. 4. Enter Caption

Brahm Ludwig (BL). Three innocent personas are also associated with the **illegal images** case: Paul Djanko (PD) and Jenny Dennings (JD). PD features as a 'guilty party' in another case which we describe later.

DD is related to PD. She is most likely his wife, but could be a sister, even mother. The exact relationship is never made clear. DD and PD share the same computer with different user accounts.

RB is involved with the possession, distribution, but not creation of illegal images. RB is a landlord. His user account contains a lot of activity relating to property searches, downloading of property brochures, and subscriptions to landlord newsletters. RB and Jenny Dennings (JD) share the same computer with different user accounts. The relationship between RB and JD is not clear, however, some learners provided unsubstantiated speculations that they are boyfriend-girlfriend.

BL's real name is Pedro. BL is involved with the possession, distribution, but not creation, of illegal images.

Figure 4 (top left) describes the case as a social network with connections highlighted in dashed lines. Personas involved in the **illegal image** case are identified with an I next to their image.

4.1.2 Illegal Images: Narrative The case study assumes that the possession, distribution, and creation of images of 'Batman' are illegal. There are two types of image. Category A images – attracting a higher custodial sentence – are real photographs of 'Batman'. Category B images – which attract less harsh sentence – are non-real cartoon/other photos of 'Batman'.

We simulated the CPS prosecution guidelines which set a test of proportionality as follows: *"prosecutors should apply a proportionate assessment to the number of images presented to a court in order to deal with these cases justly, efficiently and expeditiously ...[of]... at least 250 images at Category A"* [CPS (2018)]. We set the limit at 75 category A and 75 category B. This aimed to ensure that learners were not spending most of their time sifting through more than a thousand images. learners should aim to find more than proportionality test (maybe 100 for each category), but did not need to categorise every single image.

DD, RB, and BL have searched for and downloaded illegal images using web browsers. Images are saved in directories. Some directory names indicate a conscious effort to organise illegal images, e.g. `\Batman images`. One folder was created on a desktop.

Files have been moved, deleted, and in some cases permanently deleted from the recycle bin. A handful of extensions have been changed, in one case a directory has been hidden.

As time progresses, the personas begin emailing each other. Images are shared between personas:

- Within the email body.
- As attachments.
- As zipped attachments.
- Through WeTransfer and/or Dropbox.

In one of the emails, DD tells RB and BL about a Google extension they should install to enable faster download of images. All three have installed this Google extension and use it to download images.

Malware. In interview, all three suspects tabled the 'malware defence', arguing that malware must have caused the downloads to happen. As it happens, there is indeed malware on the system.

The researcher installed benign malware onto each computer. Benign malware is malware which is recognised by anti-malware tools as malware but is actually harmless.

Benign malware was downloaded onto our own machine in a sandbox environment (not on any of the persona machines). The file was renamed with a `.jpg` extension, for example `image.jpg`, and then hosted on a file server with several batman images. A persona accessed the file server, and downloaded the images including the malware onto his / her own computer. The persona then shared these images with others through email, WeTransfer, or Dropbox. Indeed, the personas could just have shared the file server link.

Trip Hazards. We installed several trip hazards within the case.

- Some directories were deliberately named to assess investigative rigour. One persona created a directory called `\My Colection`. Some learners either consciously or subconsciously (autocorrect) renamed this to `\My Collection`.

- The timezone on one computer was set to be Pacific Time Zone (PT). This created challenges in timeline creation.
- The presence of JD and PD provided an opportunity to create the master of all trip hazards, the investigation of an innocent persona.

4.1.3 Illegal Images: Assessment Requirements The assessment required learners to investigate the presence and provenance of illegal images, to identify other potential suspects, and to respond to the comments made during interview – one of which was the 'malware defence'.

learners should investigate email communications to determine who the source of images is. They should develop a chronology of activity including email communication and browser activity.

We expect learners to identify that there is malware on the system, and investigate the capability of the malware to determine whether the malware could have been responsible for downloading images.

We expect learners to investigate the provenance of images, how did they get onto the system? Learners might investigate zone identifiers too. We do not expect learners to use Google Lens or similar tools as this would be illegal.

4.2 Drug Trafficking

4.2.1 Drug Trafficking: Personas The following personas were associated with the **drug trafficking** case: Paul Djanko (PD), Stuart Tchaikomsiva (ST), 'Birmingham Drops' (BD) – no other details available, 'Bill' – no other details available, Petrovski – no other details available, 'Kevin' (K) – no other details available, and 'Lez' (L) – no other details available. One innocent persona was associated with the case: Denise Djanko (DD).

PD is a *drug trafficker* responsible for the trafficking of illegal drugs. illegal drugs are defined as class A (red jelly beans), B (blue), and C (green) drugs each of which attracted an appropriate custodial sentence. PD is operating a dashcam which has also been seized.

Bill is a *drugs mule.* Bill collects the drugs from 'Lez' (described later), supplies them to *drug distributors* named by PD. Bill waits for the drop to take place, and then collects the money from the *drug distributor* and pays this to PD. Typically, Bill waits on a side street close to the location of the drop. The collection, distribution, and payment is done on the same day, typically within an hour.

Petrovski (PS) is a second *drugs mule.*

ST is a *drug distributor.* ST owns a Samsung Tablet. He collects drugs from Bill and then pays Bill once a drop has been completed. The drugs drops take place on two university campuses in the Coventry area of the UK.

BD aka 'Birmingham drops' (real name unknown) is a *drug distributor* specialising in the distribution of drugs in Birmingham based university campuses. He collects drugs from Bill and then pays Bill once a drop has been completed.

'Lez' (L) is a *drug supplier.* He supplies drugs to *drug mules* specified by PD.

'Kevin' (K) provides a contact to PD for a job in Birmingham. This job is undertaken by BD.

DD is the wife of PD. She uses the same computer as PD albeit a separate log in account. JD is completely innocent of any wrong doing.

Figure 4 describes the case as a social network. Personas and their roles are identified in the diagram as follows: personas are identified with a `DM` (*drugs mule*), `DT` (*drug trafficker*), `DD` (*drug distributor*), and `DS` (*drug supplier*) next to their image.

4.2.2 The Drug Trafficking: Narrative This case involves the trafficking of drugs across the universities of Birmingham, Coventry, and Warwick[1]. PD traffics class A and class B drugs through two distributors: ST and BD. None of the distributors know each other.

PD provides detailed instructions on:

- Where to collect the drugs from on the morning of the drop.
- Where to meet the recipient. This information also includes a description of a recipient, and sometimes a campus map with the meeting point circled.
- Where to drop the proceeds at the end of the drop, i.e. the precise street name/place.

The case, as described above, is a perfectly adequate assessment which can be set as it is with PD's DFI provided to learners as the source of investigation. However, the assessment needed to be scaled up for a Master's degree with learners doing a lengthier assessment. This was achieved as follows.

At some point, ST discovers that Petrovksi can supply drugs and that he (ST) could become a drug trafficker himself. Some days later, ST is assaulted at the University of Warwick. PD has been arrested on suspicion of the assault. Law enforcement have seized his dashcam.

PD's DFI and the dashcam are provided in the assessment and learners have to investigate whether PD did indeed have an involvement in this assault. Figure 4 includes the assault element of the case with personas involved with the **assault** case identified with an `A` next to their image.

4.2.3 Trip Hazards Two trip hazards were inserted into the **drug trafficking** case. Petrovsky's name is deliberately mis-spelt by a persona in an early communication to be 'Perth Ski'. A handful of learners then referred to 'Petrovsky' as 'Perth Ski' in the end report. DD is an innocent persona in this case.

[1] This is a purely fictional scenario. The universities were selected because we needed a physical campus for the geospatial element of the task. There is no suggestion intended that any of the universities, or the learners studying therein, have a drug problem.

4.2.4 Drug Trafficking: Assessment Requirements Unlike the **illegal images** case, the **drugs** case, does not involve the investigation of much browsing or app activity, but does require the consideration of multiple suspects, the analysis of lots of email activity, the analysis of connections between multiple personas, the construction of a timeline, and the analysis of geospatial data (if we include the dashcam in the investigation).

4.3 Sample Case Statement: Assault

A case statement was issued to students as follows:

At 06:10pm on the 22nd March 2023 a warrant was executed by West Midshires Police at the home address of Paul Djanko. The arrest took place following intelligence linking Paul to an assault on Stuart Tchaikomsiva.

Two items were recovered during the arrest: a laptop and a lens-damaged dashcam. Both items have been imaged and are supplied to you by West Midshires Police as: `MMX-Djanko-PC.E01` MD5 hash: `608ec9b3ccb38f87aaee4770b1483c53`, SHA1 hash: `c5d50594bc4bdeaa77cf49d0c741e67a04ec2a59` and `MMX-1-Dashcam.E01`, MD5 hash: `a66d8d5b5d2aeea75d1c8626f21507e4`, SHA1 hash: `f19b8d4bc9afc3ef 9eb10f084cd57f540eeac87f`.

During initial interview, the Suspect stated that he:

- Does not know the victim.
- Has never met or spoken with the Victim.
- Was in Leicester on the day of the assault.

Your senior officer has asked you to examine the two exhibits and:

1. Determine whether there is any evidence to connect the Suspect with the Victim.
2. Determine whether the Suspect was in the vicinity of the assault.
3. Consider each of the statements made by the suspect in interview and identify any evidence to corroborate or refute these statements.
4. Determine and identify other suspects.
5. Report any other relevant issues.

Sample Case Statement: Illegal Images

At 06:10 on the 10th March 2023 a warrant was executed by West Midshires Police ("WM Police") at the home address of Rueben Beethomahler ("The Suspect"). The arrest took place following intelligence linking The Suspect to the distribution of illegal images of Batman, the possession, creation, and distribution of which were made illegal in 2007. 1.1 The law Subject to sufficient evidence being produced, WM Police believe they may be able to refer the case to the CPS under the Criminal Justice Act ("Batman" images) 1984 s160 Possession of photographs of Batman. Relevant sections of the Act are provided in Appendices 2. 1.2 Exhibits A laptop computer was seized during the arrest. The computer was imaged and is supplied to you by WM Police as: MMX-Beethomahler-PC.E01 (comprising 5 files in total with the extensions .E01, .E02, .E03, .E04, and .E05) The associated hashes are as follows: MD5 checksum: 7f615f9038aef7348edbca1187f2ac0c SHA1 checksum: c921ee619388fe569876749f63c03740378d8326 Additionally, a case file has been created and supplied by the case technician with all relevant ingest modules activated. 1.3 Interview During initial interview, the Suspect stated that he: • Is not 'interested in images like that'. • If these images appeared on his computer, 'they must have appeared there because the computer was hacked, or because of a virus'. • If he ever saw images like that, he would have 'deleted them straight away'. • He has never 'been in contact with anybody that may have expressed an interest in images like that'. 1.4 Task Your senior officer has asked you to examine the exhibit and: 1. Reveal any evidence that might determine whether the suspect created, was in possession of, and/or distributed illegal images. 2. Determine and identify other suspects. 3. Identify the sources/provenance of any images found. 4. Report any other relevant issues. The findings must be presented as an expert witness report, a template is provided in Appendix 3. You must not stray from the template.

5 Conclusions

We have described a six stage process which we have applied three times now to create multiple *investigatve DFIs*. The six stage process can be followed by other researchers to create multiple DFIs.

We can make several important observations/reflections on our experiences of this process.

5.1 Noise

The presence of noise is critical in an investigative scenario. Subscriptions to newsletters and our best efforts in simulating 'normal' user activity such as browsing for work/hobby/family related interests generated some noise. However, the type of noise generated was limited as we were not able, for legal/ethical reasons, to directly engage with real life entities such as companies and/or people. Nevertheless the activity we have described generated sufficient noise for our purposes.

There need to be better and more effective methods of generating noise. AI or other automated mechanisms could be used to help generate synthetic noise. However, this is not without challenges.

The biggest challenge in creating noise is maintaining technical consistency and plausibility. Technical consistency in that although one may be able to create a fake browsing history, this runs the risk of creating unplanned inconsistencies. Each event timestamp in the browsing history must be consistent with:

- Creation of cookies/other fragments created for each website when validly visited.
- The system timeline. The fake browser entries must be consistent with a period where the user was logged in.

- Event log entries must be consistent with browser activity. For example, if `Audit process tracking` has been enabled in the event log for the browser, then there should exist an entry for the start and stop of the browser in the event log.
- Entries in the registry.

5.2 Time

Creating the cases took a considerable time. Half an hour to forty minutes per day were spent to enacting the narrative.

This required considerable discipline. At the end of each day's play, several threads were often left hanging. For example, a drug distributor had promised to report back on a drop, or an illegal image distributor had promised to share photos the following day. This had to be followed up otherwise an element of credibility would be lost.

In a realistic scenario, personas would be interacting with their computer system and communicating with others throughout the day. In the present case, the activity was clumped together within half hour to forty five minute periods.

5.3 63 Cases

The six DFIs gave potential for 63 possible cases as described in Table 1. The Table describes how cases can be used individually and or combined to provide a large stock of assessment material. An example of how this can be applied is as follows:

1. Case 1 (row 1) involves the arrest of ST on suspicion of drug distribution. ST's DFI is provided. The case remit could be changed slightly to charge him with drug trafficking (whereas he is actually a distributor not a trafficker).
2. Case 2 (row 2) involves the arrest of PD on suspicion of drug trafficking. PD's DFI is provided.
3. Case 3 involves the arrest of PD and ST on drug trafficking and distribution charges. PD and ST's DFIs are provided.

The addition of the dashcam adds several more cases.

Table 1. 63 forensic cases.

Case number	BL	RB	JD	DD	PD	ST
1	0	0	0	0	0	1
2	0	0	0	0	1	0
3	0	0	0	0	1	1
4	0	0	0	1	0	0
..	..	..	..	..	..	..
63	1	1	1	1	1	1

References

Lallie, H.S., Lawson, P., Day, D.: Using digital logs to reduce academic misdemeanour by students in digital forensic assessments. J. Inf. Technol. Educ. **10**, 255–269 (2011)

Forensic, X.S.: Computer forensics : Hacking case using autopsy (2020). https://www.forensicxs.com/computer-forensics-hacking-case-using-autopsy/

Lallie, H.S.: The use of digital forensic case studies for teaching and assessment. Cybercrime Forensics Education and Training, pp. 1–9 (2010)

Horsman, G., Lyle, J.R.: Dataset construction challenges for digital forensics. Forensic Sci. Int. Dig. Invest. **38**, 301264 (2021)

Grajeda, C., Breitinger, F., Baggili, I.: Availability of datasets for digital forensics-and what is missing. Digit. Investig. **22**, S94–S105 (2017)

Breitinger, F., Jotterand, A.: Sharing datasets for digital forensic: a novel taxonomy and legal concerns. Forensic Sci. Int. Digital Invest. **45**, 301562 (2023)

NIST, Computer forensic reference dataset portal (2023). https://cfreds.nist.gov/

Garfinkel, R., Farrell, D.: Bringing science to digital forensics with standardized forensic corpora (2009). https://digitalcorpora.org/corpora/disk-images/

Carrier, B.: Digital forensics tool testing images (2010). https://dftt.sourceforge.net/

CPS. Indecent and prohibited images of children (2018). https://www.cps.gov.uk/legal-guidance/indecent-and-prohibited-images-children

Driving the Future of Digital Forensics

Rachael Medhurst and Beth Jenkins(✉)

University of South Wales, Newport, UK
{rachael.medhurst,beth.jenkins}@southwales.ac.uk

Abstract. Enhancing the pedagogy of digital forensics is crucial for academia and industry. Digital Forensics is an industry that is ever-changing and, as such, academia needs to adapt to stay current. Developing Digital Forensic University courses are required to bridge the gap between academia and industry, thus producing industry ready graduates. Traditionally, teaching within university's has been lecture-led, often sidelining the importance of practical learning. Experiential learning theory states that "experiential learning is the process of learning by doing. By engaging students in hands-on experiences and reflections" (Kent, 2024). Students are participating in practical activities, this increases transferrable skills such as communication, teamwork, and problem solving. To achieve this, enhanced practical learning is required, this increases industry visibility and authentic learning experiences. This can be done through different methods such as collaboration with industry, research, courses/training, conferences, real-world case scenarios, laboratory learning environments, and authentic learning experiences. With the incorporation of these practices, this helps to develop students' knowledge and develop industry ready graduates to target the industry's aim of fulfilling the roles of the rising digital forensics opportunities available. This is a continual process that requires consistent development but will in turn assist academia and industry collaboratively.

1 Introduction

Developing teaching capabilities is a multi-faceted approach, that requires continuous development to ensure that students are developing and learning in the ever-evolving field of digital forensics efficiently within Higher Education (HE); which can be embedded within industry. This has the ability to enhance the industry further. This is achieved by bridging the gap between academia and industry to provide industry-ready students upon graduation, enhancing employability and the economic impact of this field. Paul Jacques has quoted within the Police Professional article in 2022 that "forces are overwhelmed by digital forensics with a backlog of 25,000 devices waiting to be examined". This is a common theme within the digital forensics industry and continuing to rise with the Police Digital Service (2024) stating that "90% of reported crime now has a digital element". This trend has emerged as cyber-enabled and cyber-related crimes have become common methods for committing criminal activities, necessitating ongoing development and training of personnel within the field.

P. Legg et al. (Eds.): CSE 2024, LNNS 1213, pp. 39–53, 2024.
https://doi.org/10.1007/978-3-031-77524-6_3

To support the expanding field of digital forensics and the increasing demand for these skills, universities must enhance their teaching capabilities to effectively teach and develop personnel in this area. While this endeavour can be both challenging and rewarding, it requires incorporating several key components, which will be discussed further within this section. These components include learning methods, pedagogical development through preparation, teaching content, formative and summative assessments, academic feedback, student and industry feedback, and ongoing continuous development. With these elements incorporated, and actioned, this enhances educational outcomes, meets the industry needs while increasing graduates' opportunities and professional development.

2 Learning Methods Within Higher Education

Higher Education (HE) is a well-established sector that develops individuals to their associated industries, however, traditionally within HE the approach has been more established on "rote" learning where students are required to memorise content without the need to understand the practical application that would otherwise be required for industry. The content is often delivered in traditional methods such as textbooks and board-based instruction provided by the academic, however, this can lack the deeper comprehension that is required to develop knowledge within the associated field. This is particularly evident within a niche and specialist field of digital forensics that requires a deep level of practical understanding and knowledge to efficiently apply within industry.

With the previous emphasis on theoretical teaching practices in HE this has resulted in graduates requiring further development in their job roles to effectively perform digital forensic investigations or related work in the field. This challenge is exacerbated by the rapid increase in digital devices requiring investigation, the complexity of devices examined, along with the requirement to increase staff numbers to handle such investigations. The Police Foundation highlighted in 2021 that "staff numbers would have to increase by between a third and a half to meet current demand". To address this issue and the continual development of teaching practices, HE requires authentic learning experiences and therefore, HE should incorporate teaching methods such as "experiential" learning. This enables students to learn in a practical manner that can be applied within the industry to target these challenges. However, all academics should be aware that this is a continuous process to ensure that they persist to enhance learning capabilities within the field.

"Experiential learning" is when the student learns by 'doing', which consists of four stages, as explained by Kolb theory, which is experiencing, reflecting, thinking and acting. With these four elements incorporated, this allows the students to bridge the theory and practice that is aligned with real-world situations. Raymond Flores et al. completed a study about 'using experiential learning to teach and learn digital forensics resulting educator and student perspectives' which found that hands-on real-world experiences have been found to be beneficial'.

For academics to be able to achieve effective experiential learning where Raymond Flores et al. has stated that 'students engage in active hands-on, authentic learning activities through which they apply their knowledge and skills in real-world contexts'. To be able to achieve this, there are various aspects of the teaching process that should be reviewed and amended where possible. This includes preparation, teaching content, formative and summative assessments, feedback, but also the tools and facilities required to change teaching practices to provide increased authentic experiences. Therefore, the next section is going to delve deeper into these areas to provide further insight into the processes already actioned. However, it is vital to consider that this is a continuous process requiring feedback, reflection, continuous monitoring and flexibility.

2.1 Preparation

Effective teaching preparation is essential and fundamental for academics to deliver high-quality material required to meet the diverse needs of students while targeting the subject field efficiently. This includes several aspects, understanding the subject matter, curriculum development, lesson planning, pedagogical strategies, classroom management, assessment, feedback, professional development and reflective practice. These processes are fundamental. However, often the correlation between industry and academia can be misaligned, or academia are more theoretical based, lacking that depth required within digital forensics. To be able to do this effectively, several elements can be embedded during this preparation stage including, focus groups, facilities including classrooms/laboratory space, industry engagement, equipment and training requirements. Each of these preparation aspects are going to be discussed in further depth to provide clarity on enhancing teaching practices to target the experiential development of students. The aim of embedding these practices into the preparation stage will develop more graduates that are capable of working effectively within the digital forensic industry and assisting in the expertise shortfall currently being witnessed.

2.2 Industry and Academia Focus Groups

Before the academic prepares teaching materials, a deeper understanding of the digital forensics industry is necessary - input from industry professionals is invaluable. This can be achieved through focus group sessions where the relationship of academia and industry is developed and aligned. The University of Cambridge has stated that "a focus group discussion is a unique method of qualitative research that involves discussing a specific set of issues with a pre-determined group of people" (Cambridge, 2021). This allows a dedicated time for academia to share the course and module specifications, allowing industry to have further input before the changes are implemented. To be able to achieve this, this is a step-by-step process which would form part of several meetings with industry and academia, which includes the following (Fig. 1):

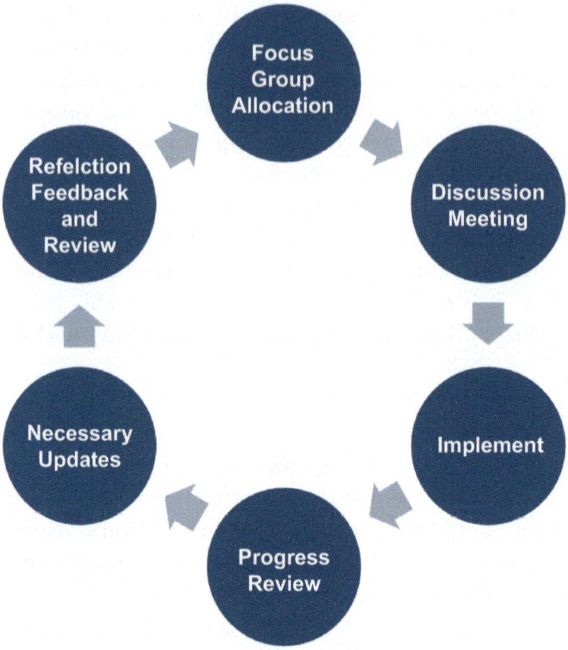

Fig. 1. Focus Group Cycle

2.2.1 Focus Group Allocation As the Digital Forensic courses can incorporate several elements of the industry within their courses and related modules, a variety of skillsets and specialisms within digital forensics is required to form the focus groups. This could include specialisms such as Digital Forensic Investigators, Quality Managers, Digital Forensic Managers and Digital Forensics Incident Response personnel. This allows for different employees/companies to form part of this focus group which is beneficial to provide a wider insight into the field and elements of authentic learning that can be embedded within the courses/modules where appropriate. While this focus group may be a challenge to form, the overall benefits to both industry and academia cannot be understated. Often industry and academia identify the benefits and willing to collaborate to target the overall aim of authentic experiences that would benefit the students, academia and industry.

2.2.2 Discussion Meeting Once the focus group has been formed, a premeeting should be conducted at a suitable date and time. During this meeting, the initial introductions can be made, along with explanations of the focus group plan, structure, the overall course and required expectations. This allows industry to see the current digital forensic provision, along with the proposed changes that industry can review and provide feedforward information to align the content to industry expectations, industry feedback and input, ultimately with the goal of providing students the opportunity to receive authentic experiences that can be embedded within industry upon graduation. The proposed changes from industry are noted during this meeting, and the academic team continues with the minor or major modifications (dependant on the change) within the associated module, revalidation of modules, or validation of new modules, along with consideration of additional resources required. Resources could be human factors, software, hardware, and teaching spaces.

2.2.3 Implement After the discussion meeting concludes, the necessary expectations and suggested initiatives to integrate into the course and modules will be identified. The next logical step is for the academic to implement these changes by modifying existing modules, introducing new modules, update module specifications, developing case scenarios, and determining equipment requirements. Following these actions, a progress review will be needed to assess the feasibility of implementing these changes.

2.2.4 Progress Review After the pre-meeting has been completed, continual progress reviews, and feedback to industry is required. This allows time allocation to reflect on those changes being implemented, challenges encountered, and potential solutions that could be incorporated. Once this progress review has been completed, then the continuation of preparation is required with the updated solutions and ideas.

2.2.5 Necessary Updates After the previous meeting of the progress review where challenges and potential solutions are discussed, this meeting is about discussing those challenges and updates of the proposed solutions to establish if the challenges have been overcome. If there are any further issues that have arisen, this is the opportunity to discuss further, ready to implement any actions where necessary. If further challenges have arisen, then it is good practice to go back to the pre-meeting and explain the expectations, goals and aims to establish if the same outcome can be achieved in a different manner, then go through the implementation process again.

2.2.6 Reflection, Feedback and Review Once all the updates have been actioned and placed within the associated module, the importance of reflection, feedback, and review cannot be underestimated. This can be completed at different stages of the implementation, including:

1. Initially during the meetings and progress reviews.
2. Team review of amendments to the course/module specifications.
3. Approval from students and External Examiners (EEs) to action changes.
4. At the Faculty Quality Assurance Committee (FQAC) before the changes can be approved or denied based upon the University's policies and procedures.

5. The academic team should conduct reflections after the change has been implemented within the teaching practice and students have begun to engage with the material.
6. Anonymous feedback opportunities for students to provide feedback, both positive and constructive.
7. Feedback opportunities from industry about the changes implemented, and steps moving forward.
8. Student and Staff Course Liaison Groups (SSCLG) to discuss the changes on the course, and the impact of these changes implemented.
9. Finally, when finishing a university course, students complete an NSS (National Student Survey) which is a survey for UK students to provide their opinions on the quality of the course. This should be monitored and compared to the previous years to determine if the overall quality of the course has improved.

Therefore, there are multiple opportunities for the academic, university and industry to reflect on these changes and gather feedback. While gathering feedback is essential, the review of the gathered data is required to be able to determine good practice, challenges and amendments for the future years. This is a continual cycle, and therefore, should not stop at any one stage.

3 Facilities

To effectively implement experiential learning and offer students an authentic experience, it is essential to consider the learning environment and facilities available. Traditionally, lecture halls have been used to deliver information about the related subject area, however, this limits the potential for networking, practical learning and a sense of belonging for students whose desires are to have a professional career within digital forensics. Anthony Herrington from the University of Wollongoing, Australia has stated that 'the approach taken by many teachers in universities today is simply a result of the way they were taught. They are perpetuating a tradition of formal university teaching has ignored the substantial insights gained from more recent theory and research into the way people learn'. Therefore, the transition from traditional lecture halls to a digital forensics lab or similar is beneficial. However, further to this the incorporation of the necessary equipment and tools is required. To be able to achieve this, the incorporation of a computer laboratory would be required along with the industry-based tools, both hardware and software, providing students with an authentic experience and embedding themselves within the role of a digital forensic professional witnessed within industry. By emulating a digital forensics environment this improves the students learning experience and provides a different thought process from a student to a professional by conducting scenario-based case work, real-world situations and embedding the experience of the specialised role within academia.

To be able to achieve this, case scenarios need to be developed that are authentic to the nature of the digital forensics industry. There should be various aspects considered when developing the case scenarios, including the level of teaching, pedagogical developments, types of devices, operating systems, anti-forensics techniques implementation, etc. Depending on the level, there should be different learning outcomes. For example, level 4 may be more focused on the digital forensic process adhering to all principles and procedures on computer systems, and other devices such as USBs, as opposed to the identification of evidence itself, and any use of anti-forensic techniques. Whereas, when developing through the higher levels, there should be more consideration of additional devices that are more complex to investigate, such as mobile devices, IoT devices, anti-forensic techniques embedded into the case including encryption, data wiping tools etc. that would assess the digital forensic process, along with the investigation methods and evidence identified.

4 Industry Engagement

While the focus groups allow industry to have an input into the content, delivery and assessments of the course and associated modules, the requirements for industry engagement throughout the teaching and delivery of such content cannot be understated. Dr Rose Leahy, et al., has stated that the role of an engaged, entrepreneurial university is to act as a co-creator of knowledge, informed by, and in collaboration with, external organisations, and to ensure that graduates emerge with the academic excellence, practitioner knowledge and entrepreneurial skills and capabilities which ensure that they are well-placed to identify or create employment opportunities and to maintain employment'. Industry engagement allows for authentic learning to be fully embedded into the curriculum. By creating this dialogue between academia and industry, all parties benefit. Industry can have input into the development of teaching material and assessments, and academia develop their understanding and changes of the industry, which in turn further aligns industry and academia together. This ensures that material remains relevant and equips students with the necessary skills to handle case work in their careers.

5 Continual Professional Development (CPD)

While implementing all these changes are vital, it is important that the academic understands and can embrace practical teaching and enforce industry-based knowledge within the learning environment. Therefore, the role of Continual Professional Development (CPD) is going to be fundamental. Once the CPD has been completed and further developed, the overarching goal of industry-ready graduates is closer to being realised. Therefore, the investment into elements such as facilities, industry partners and development of academic teaching practices is a necessity to ensure that the overall aim of the course is achieved. This should be done on a frequent basis due to the changes within the industry. This can be achieved in various manners, including information exchanges between academia and industry, training courses, knowledge development through forums, webinars, along with research, presenting at conferences and networking opportunities. With the incorporation and support of such activities, this can all be used to enhance learning experiences for the students and enabling the sharing of knowledge to develop graduate-ready individuals that can support the industry efficiently.

6 Teaching Content

Traditionally, teaching has been more focused on presentations of information as opposed to embedding that learning. However, the presentation of such information cannot be understated, and does have a place within the learning environment. Therefore, this is incorporated within the learning process, this is completed as a 'flipped' classroom process. Advance HE has stated that flipped learning is 'a pedagogical approach in which the conventional notion of classroom-based learning is inverted; students are introduced to the learning material before the class, then within the classroom time being used to deepen understanding through discussion with peers and problem-solving activities facilitated by teachers'. This allows time spent with the academic to be more directly focused on authentic and practical learning experiences for students to develop upon their theoretical understanding. This does require students to be autonomous learners where they take control of their own learning experiences and development. The 'flipped' classroom approach does not just mean presentations of information, this could be recorded presentations with discussions, quizzes, research tasks, engagement with webinars, etc. and incorporate technology enhanced learning (TEL) and blended learning which allows for multiple methods to deliver learning, could be face to face or online activities.

Following the 'flipped' classroom approach while the students are within the university environment, there may be a quick recap of this material and opportunity for questions/discussions, but the onus is on the students to have already reviewed the resources. This allows the on-campus teaching to concentrate on challenge-based, experiential learning and practical workshops, helping to solidify theoretical knowledge into active, practical understanding that can be applied in the industry. This can be further assisted with the use of 'formative' assessments.

6.1 Formative Assessment

Formative assessments play a pivotal role in education by facilitating the development of students' knowledge and skills while identifying areas that require improvement. This type of assessment focuses on non-assessed exercises designed to enhance students' strengths and allow them to address their weaknesses. For instance, in the field of digital forensics, formative assessments can be effectively implemented through investigation scenarios. In these scenarios, students are assigned cases involving various digital devices, such as hard drives, USBs, and mobile devices. Acting as principal investigators, students engage in contemporaneous notetaking and work through forensic images to answer a set of questions, ultimately presenting their findings to the class. This approach enables students to 'learn through doing', thereby honing their investigative skills. Formative assessments are also instrumental in developing students' transferable skills. A notable example is the creation of a 'murder mystery' exercise. This exercise centres on a fictional murder case with multiple plausible suspects, deliberately excluding digital forensic elements to emphasise communication, report- writing, and presentation skills. Students collaborate in groups to build a case and identify a suspect based on the provided evidence. Initially, each group receives two sets of evidence outlining the scenario and introducing the characters, each with a potential motive for the murder. The groups can request only two additional sets of documents from a list of available evidence, challenging them to make strategic decisions and draw well - supported conclusions. The exercise culminates with the groups presenting their findings, including the identified suspect and motive, to the class. This activity not only fosters teamwork and analytical thinking but also simulates the pressure and deadlines found in industry environments. By removing the digital forensic component, the exercise compels students to focus on developing their transferable skills, essential for success in professional settings.

Formative assessments are also utilised in class through the investigations that students complete. They are given scenarios and provided with various forms of evidence, such as conventional hard drives, SSDs, USB sticks, and mobile devices. The students receive a set of questions designed to guide them through each category of the forensic image, for instance, by examining media files, text messages, and emails. These questions are structured to encourage the application of different digital forensic techniques, such as using search functions and filtering, to find the answers. This approach not only aids in reinforcing their understanding of forensic processes but also promotes the development of practical skills essential for digital investigations.

Formative assessments are a valuable tool in digital forensic education. It supports the development of both technical and transferable skills, preparing students for real-world challenges by promoting active learning and critical thinking.

6.2 Formative Assessment Feedback

While formative assessment does not directly feed into the overall grade of the associated module/course, there is a requirement for feedback to be provided to allow students to develop understanding and knowledge ready for the summative assessments required. Stanford Teaching Commons has stated that "Formative assessments are usually meant to measure learning in order to provide feedback. This kind of formative feedback is given throughout the duration of the course and is given in a spirit of growth and improvement". This is essential to allow students to understand areas of growth and what areas that require improvement. Feedback can be provided to students in various formats, this could be in person, written feedback, audio/video based, but additionally peer feedback can be beneficial to learn from each other. While this may take sufficient time, this is a crucial part of the learning process.

7 Summative Assessment

Experiential learning in digital forensics can be effectively facilitated through the use of summative assessment. This can be integrated into modules by incorporating an end-to-end investigation as a form of assessment. The end-to-end investigation process immerses students in all aspects of a digital forensic investigation. This includes the collection of evidence from a scene, interaction with occupants and addressing their queries, conducting a thorough investigation, drafting a court-worthy expert witness report, and undergoing cross-examination in a court setting. This comprehensive assessment enables students to conduct investigations collaboratively, thus enhancing their investigative skills.

During this end-to-end investigation, students are required to attend a simulated crime scene house, where they are met by a homeowner who may or may not be the person of interest in the investigation. They must navigate the scene and collect digital evidence. In this assessment, the homeowner will deliberately act disruptively, encouraging students to develop their transferable skills and, crucially, to understand different perspectives. While the students may be performing a familiar task, the homeowner may find the situation highly stressful. Feedback from this assessment has revealed that students did not adequately consider the perspective of the homeowner. Prior to completing the crime scene assessment, students reported difficulty in perceiving the homeowner or 'suspect' as a 'real person.' Recognising this perspective is crucial in their development as investigators, emphasising the importance of empathy and understanding in forensic investigations.

The students are then tasked with collecting as many pieces of digital evidence as possible but are permitted to seize only three devices. This limitation encourages students to develop their understanding of sampling. According to the forensic regulation ISO/IEC 17025 that is used for digital forensics states that sampling involves defining a procedure to select a smaller subgroup of items from a larger group. In the context of digital forensics, the criteria for selection might include the condition of the items (e.g., damaged or too old to be examined) or the availability of laboratory resources. This exercise aims to enhance students' practical skills and their ability to make informed decisions under pressure.

Once the students have seized evidence from the crime scene, they will utilise their selected devices as the foundation of their investigation. They will analyse each piece of evidence and subsequently create an expert witness report to be presented in court. The students are required to submit a case folder, which includes their expert witness report, contemporaneous notes, and any other relevant documents, such as crime scene photographs. In court, the students will present their findings and answer questions pertaining to any documentation they have submitted. The court will consist of a prosecution barrister, a defence barrister, a judge, and a public gallery. This setting encourages students to critically reflect on their work, fostering a deeper understanding of the requirements for an expert witness report.

These real-world scenarios have been created to provide students with an authentic experience that can be implemented within industry upon graduation. These develop technical skills, and related digital forensic skills such as expert witness report writing, court room skills, but additionally a large range of transferable skills such as communication, organisational skills and team working. These are all vital ready for industry.

7.1 Summative Assessment Feedback

Previously, formative assessment feedback should have been provided to students to allow further understanding of gaps and areas of growth. Stanford Teaching Commons has stated, regarding summative assessment, that "these assessments will address how well the student is able to synthesise and connect the elements of learning from the entirety of the course into a holistic understanding and provide an opportunity to provide rich summative feedback". This is particularly useful when there is a module with two assessments, and the student must complete both assessments to target the learning objectives and develop knowledge of the module area. That feedback will be vital to enhance quality of the second assessment by the student understanding what they need to develop.

Summative assessment feedback can be provided in a variety of ways, including verbal, written, audio/video, etc. However, the incorporation of marking rubrics has made this clearer for the student to understand the key areas of good practice, and for their development. While the academic uses this rubric to conduct their marking, it is beneficial to provide the rubric to the student when they are conducting the summative assessment. This will allow students to understand what the academics are marking against and to target the key areas they want to achieve. When the rubric is used to provide feedback, this provides a clear and transparent method of providing feedback. However, there are other benefits including consistency across marking of students and assessments, as well as being more time efficient for the academic. While this is a chart/table format, additional written feedback can be provided to provide further clarity and guidance where needed for the student to develop further.

8 Student and Industry Feedback

While implementing such changes can be seen as a positive move, feedback on these changes from several parties involved is required, this includes the academic's own reflective practice, student and industry feedback. This enables continuous reflective practice and development to ensure that the overall quality of teaching is constantly enhanced, and the delivery of industry-ready graduates are produced.

8.1 Student Feedback

The University of Greenwich (2024) has stated that student 'feedback influences how students feel about their course, their performance and themselves. It is important then to structure feedback in such a way as to maintain or increase students' motivation and to encourage them to focus on learning goals'. There are several opportunities for students to provide feedback to the university throughout the academic year, and at the end of their studies which allows the academic team to review, reflect, and implement changes where necessary. This feedback should incorporate several elements of the changes including, positive, improvements, recognition, and constructive criticism.

This can be done anonymously through a feedback platform. This allows students to provide feedback at several points throughout the academic year, this can be done for course and module level. This feedback is completely anonymous and reviewed by the associated teaching team, and senior management to support actions that could be implemented. Another method is the progression meetings, which are conducted throughout the academic year with the teaching team and the class representatives from the related course to provide feedback to the academics about their student experience. This is a face-to-face meeting, and not anonymous, but students are encouraged to speak to their course representative before the meeting to allow the course representative to give a full insight into the running of the course. During the meeting, all feedback is documented, allowing the feedback and any comments to be shared with the entire course. The last element is the NSS (National Student Survey) that is completed by final year students as they finish their course, this allows the university to review, plan, and implement changes that could continuously improve the experience for future cohorts of students.

8.2 Industry Feedback

As the industry parties have been involved in the entire process, it is important that there is feedback provided from industry about the changes within the modules, assessments, workshops and student experiences to ensure that the original plans have been implemented effectively, and continuous improvements are made. This feedback is conducted through various forms, such as written, and verbal feedback to allow developments to be continuously made.

Industry partners also value the opportunity to provide feedback on the course. As previously mentioned, these partners can be involved throughout the entire process of course creation and module development. This collaboration enables industry partners to offer input on the materials they believe should be included in the modules. Consequently, universities can offer an authentic learning experience for students, ensuring that graduates possess the requisite knowledge and skills to succeed in the industry.

Industry feedback provides a unique frame of reference for the course, in that the modules can be continuously updated to ensure they are relevant to the work that is being undertaken in the industry. This is crucial to the development of the course. From a student perspective, the students are aware of the implementation of industry feedback and are more receptive to any changes being made within the course when they realise that it is being implemented through the industry feedback. Students also realise that the industry feedback is critical to ensure that what they are learning will enable them to 'hit the ground running' in their careers.

9 Challenges for Academics?

While all these changes and considerations have been beneficial within academia, this does not come without any challenges. While challenges are expected, there are several considerations of embedding such a practical nature within academia. These challenges include preparation, constant developments with technology and digital forensics, embedding this knowledge and expertise within practical workshops, and ongoing communication with industry. Therefore, it is essential that continuous developments are made.

9.1 Embedding Solutions

While there are several challenges to ensure that academia maintains 'applied' and practical teaching methods are implemented to continue to develop industry ready graduates. However, there are solutions that can be implemented to assist in continuing this enhanced teaching approach. This includes continual CPD, encompassing training within the subject area and teaching practices. As digital forensics is a continually developing area and the change within technology is very fast paced, it is important that the academic is continually developing knowledge of the subject area to ensure that the most current developments to the industry are being communicated effectively to the students that will be fulfilling the roles within the industry. While the development of knowledge within the subject area is vital, the continuous improvements to teaching practices are equally as important to ensure that the subject area is taught effectively.

Other methods of CPD includes continuous research, conferences and networking events. This will help to understand the developing changes within the subject area, teaching practices, allow academics to present their research at conferences, attend conferences to learn from others and continue to build and maintain relationships with industry partners. With all these aspects built together, this helps to ensure that the academic is well equipped to teach the next generation of digital forensic experts through a collaborative effort.

An important aspect of all these changes is the incorporation of industry, therefore maintaining and enhancing the relationships between academia and industry is vital. This can be a challenge due to the daily pressures of each job role, whether industry or academia. However, through collaboration this can alleviate issues with both sectors coming together to target teaching, development and job opportunities to ensure that students are working on real-world scenarios, enhancing knowledge, students assisting industry with real-world problems that can be embedded into the field, and developing graduates to fulfil the continuing growing field. This is an equally important relationship for both academia and industry.

10 Conclusion

The aim of this section of the book is to highlight the relationships between academia and industry to enhance teaching practices to ensure that students are industry-ready upon graduation to target the forever growing and changing field of digital forensics, regardless of their chosen sector of the field. Whether this is quality management, investigations, incident response and many others. While there can be challenges with developing and maintaining relationships, continuously improving knowledge of subject area, teaching practices, and considering resources that are needed to have an 'applied' learning environment. There are a range of solutions that can be embedded to overcome challenges which include proactive actions such as continuous development of subject area, teaching practices, networking with industry, research and conferences. This will all assist in the development of students to be industry-ready graduates. While there have been a significant number of changes within academia to further embed knowledge of digital forensics, this a continuous process. This requires maintenance as further improvements are needed to ensure that industry relationships and enhancement to teaching practices are developed. With these areas maintained and developed, this will fulfil the gap within the industry previously highlighted and further enhance the capabilities and resources available within the industry as the field continues to evolve.

References

AdvanceHE. Flipped Learning (2024). https://www.advance-he.ac.uk/knowledge-hub/flipped-learning#:~:text=In%20the%20flipped%20classroom%20students,this%20to%20further%20their%20learning. Accessed 15 Jul 2024

Cambridge. Introduction to Focus Group Research, Cambridge : Cambridge University Press. Commons, S. T., 2024. Formative Assessment Feedback (2021). https://teachingcommons.stanford.edu/teaching-guides/foundations-course-design/feedback-and-assessment/formative-assessment-and-feedback. Accessed 14 Jul 2024

Flores, R.: Using Experiential Learning to Teach and Learn Digital Forensics: Educator and Student Perspectives. Elsevier, p. 6 (2021)

Foundation, P.: Unleashing The Value of Digital Forensics (2021). https://www.police-foundation.org.uk/publication/unleashing-the-value-of-digital-forensics/. Accessed 18 Jun 2024

Greenwich, U. o.: Principles of Good Feedback (2024). https://www.gre.ac.uk/learning-teaching/assessment/feedback/principles#:~:text=Feedback%20influences%20how%20students%20feel,(i.e%20passing%20the%20test). Accessed 14 Jun 2024

Herrington, A.: What is an Authentic Learning Environment. IGI Global, Wollongong (2008)

Jacques, P.: Police Professional (2022). https://policeprofessional.com/news/forces-overwhelmed-by-digital-forensics-with-backlog-of-25000-devices-waiting-to-be-examined/. Accessed 17 Jun 2024

Kent. What is Experiential Learning and Why Is It Important? (2024). https://www.kent.edu/community/what-experiential-learning-and-why-it-important#:~:text=Experiential%20Learning%20is%20the%20process,classroom%20to%20real%2Dworld%20situations

Leahy, D.R.: Structures in Higher Education that Support Industry Engagement. Society for Research into Higher Education, p. 1 (2024)

Service, P.D.: The Big Picture (2024). https://pds.police.uk/national-policing-digital-strategy-2020/national-policing-digital-strategy-2020-2030/the-big-picture/. Accessed 17 Jun 2024

University, K.S.: What is Experiential Learning and Why is it imporant? (2024). https://www.kent.edu/community/what-experiential-learning-and-why-it-important. Accessed 30 May 2024

Hacktivity Cyber Security Labs: A Platform for Cyber Security Education with Randomised Challenges, Virtualisation Infrastructure Management, and CyBOK Integration

Z. Cliffe Schreuders(✉) and Thomas Shaw

Leeds Beckett University, Leeds, UK
{c.schreuders,thomas.shaw}@leedsbeckett.ac.uk

Abstract. We present Hacktivity Cyber Security Labs, a novel technical framework and hosted lab infrastructure platform. Hacktivity uniquely addresses limitations in traditional cyber security education methods, particularly through extensive incorporation of randomisation of cyber security and hacking challenges, integration of automated chatbots, provisioning and managing of VMs and datacenter clusters, and integration with the Cyber Security Body of Knowledge (CyBOK) for curriculum alignment and individualised tracking of knowledge and experience. In this paper we present the technical design details of Hacktivity and our open-source backend framework, SecGen, and reflect on our experience of integrating it into security courses.

By leveraging the SecGen framework for automatic problem generation, Hacktivity provides randomised challenges and hands-on tasks, enriching experiential learning for students. The close integration with CyBOK enables students to better understand the alignment of modules with knowledge areas and monitor and reflect on their learning progress. We present our experience of the platform's effectiveness in teaching a wide range of technical cyber security topics. This paper introduces our innovative approach to teaching computer security, leveraging SecGen, an open-source software, multiple virtualisation data centres, and Hacktivity, a new virtualisation management and virtual learning environment.

Keywords: Cyber security education · capture the flag · lab infrastructure · automatic problem generation (APG) · CyBOK (Cyber Security Body of Knowledge) · experiential learning · virtualisation infrastructure management

1 Introduction

1.1 Overview of Cyber Security Education Challenges

Hands-on experiential learning, which involves realistic and varied tasks, can be used in cyber security education to provide engaging and meaningful learning experiences. Virtualisation provides a means of providing environments and

P. Legg et al. (Eds.): CSE 2024, LNNS 1213, pp. 54–74, 2024.
https://doi.org/10.1007/978-3-031-77524-6_4

systems for experiential cyber security learning. Capture the Flag (CTF) challenges are a form of cyber security exercise where participants attempt to solve security-related challenges to capture "flags", which are revealed when the challenges are solved. CTFs can enable a hands-on and engaging way to learn and practise cyber security skills. Despite this potential, the development of CTFs and lab environment platforms face challenges.

Traditional methods of cyber security education often rely heavily on static challenges and tasks, which can lead to issues such as academic dishonesty, including plagiarism and collusion. Hosting challenge VMs remotely is crucial to prevent students from cheating by analysing disk files, rather than by arriving at solutions via intended means. Additionally, defensive and investigative tasks can be less suited to the typical capture the flag (CTF) approaches commonly used in cyber security training. These limitations underscore the need for more dynamic and immersive educational experiences. Furthermore, mapping content to learning frameworks and taxonomies has the potential to help educators develop and align curriculum, ensuring that learning outcomes are clearly defined and that educational activities are structured to meet these goals.

1.2 Introduction to Hacktivity Cyber Security Labs

To address these challenges, we present Hacktivity Cyber Security Labs, a novel technical framework and hosted lab infrastructure platform. Hacktivity is designed to overcome the limitations of traditional cyber security education methods through several innovative features. Our platform incorporates extensive randomisation of cyber security and hacking challenges, the use of automated chatbots for dynamic interaction and assessment, and the provisioning and management of virtual machines (VMs) within datacenter clusters.

Hacktivity aims to provide a user-friendly VLE lab portal to students and a course/event management system to academics.

Hacktivity provides a unique set of features not addressed by any previous approaches:

- Randomised challenges, where the technical challenges can be slightly or entirely different, mitigating collusion and cheating.
- A virtual learning environment (VLE) providing students with learning resources, and organisation around learning modules and also extra-curricular self-directed and competitive challenges ("hack boxes").
- A full-featured remotely accessible lab environment, including a variety of Linux and Windows systems, providing a safe space for conducting real attacks and security tasks, within isolated networks.
- Automated marking with various scoring systems, including providing incentives for engagement, timed tests, and late penalties.
- Gamification features, including challenge completion streaks, scoreboards, and dashboards showing students their progress, running grades, results and engagement statistics.

- A huge curriculum of content mapped to CyBOK (150+ labs and multi-step CTFs) – which can help with planning and accreditation efforts.
- A virtualisation management platform, with automated provisioning of VMs, building and managing pools of sets of VMs, so academic staff don't need to spend time managing virtualisation.
- Support for individual and team challenges.
- CyBOK visualisations including module alignment, and student knowledge tracking.
- Full desktop virtualisation, with complete work environments provided for each topic (no need for students to maintain VPNs, or local VMs).
- Based on a core value of open innovation, with content creation collaboration via GitHub.

Hacktivity integrates closely with SecGen, our open-source backend framework for Automatic Problem Generation (APG). SecGen enables the flexible specification of randomised configurations of networks of VMs, including potentially vulnerable software and configurations. This approach ensures that each student's learning experience is unique, reducing the risk of academic dishonesty and better preparing students for real-world cyber security scenarios.

Most topics within SecGen include multiple CTF challenges that are automatically marked by Hacktivity, providing instant feedback and enhancing the learning experience through continual assessment. Furthermore, many SecGen scenarios feature an interactive chatbot adversary, Hackerbot, which automates the marking of tasks and provides instant feedback. This enhances the learning experience by offering a dynamic and engaging way for students to interact with the challenges.

Hacktivity and SecGen integrate with the Cybersecurity Body of Knowledge (CyBOK), allowing for curriculum alignment and individualised tracking of students' knowledge and experience. This integration enables students to visualise their progress and reflect on what they have learned.

2 Background and Related Work

2.1 Traditional Cyber Security Training Methods

Traditional methods of cyber security education rely on static challenges and tasks. These approaches, while beneficial for introducing fundamental concepts, are prone to academic dishonesty, such as plagiarism and collusion, because solutions can be easily shared among students. Moreover, these static tasks often do not adequately simulate the variability of real-world cyber attacks, limiting their pedagogical effectiveness.

Capture the Flag (CTF) competitions are a prominent method used to engage students and professionals in cyber security training. These competitions involve solving security challenges to collect "flags," offering participants an opportunity to test and improve their skills. CTFs originated from the DEF CON hacker conference and have since become common activities at cyber security conferences and online platforms. Notable examples include PicoCTF for high school students [2] and CSAW CTF for university students [4].

CTF scenarios have demonstrated substantial pedagogical utility within higher education (HE). They have been successfully adapted for lab exercises, class competitions, and extracurricular activities. However, the static nature of traditional CTF challenges presents significant limitations when used in educational assessments. These challenges are expensive and time-consuming to create [8], and their reuse can lead to widespread solution sharing, reducing their effectiveness [1].

Automatic Problem Generation (APG) systems have been explored in various studies to enhance cyber security education by providing dynamically generated challenges. Examples include AutoCTF [8] and MetaCTF [6], which allow for the automatic generation of CTF challenges. These systems improve engagement and learning outcomes by reducing opportunities for cheating and fostering critical thinking and problem-solving skills.

2.2 Existing CTF and Training Platforms

Numerous frameworks and platforms have been developed to facilitate the hosting of CTF competitions and cyber security training. These include CTF scoring platforms such as CTFd [3] and FBCTF [5] and online training platforms such as HackTheBox [7], TryHackMe [15], and Immersive Labs [9]. While these platforms provide valuable resources for cyber security training, they often fall short in addressing the needs of educational institutions.

These platforms take the traditional approach of hosting static challenges, which are often solved and publicly shared, reducing their effectiveness for repeated use. In many cases even the flag keys are static meaning that students can share the flags without needing to even complete the challenges themselves. Creating new challenges is typically a manual process, of cloning VMs, and installing software and manually coding and configuring challenges. The lack of randomisation and dynamic content makes these platforms less effective for assessments and skill development. These challenges are not interesting to repeat as the solutions do not evolve or change on repeated attempts. Furthermore, none of these platforms track learning in terms of CyBOK.

3 Technical Framework of Hacktivity

3.1 Introduction to SecGen and Hacktivity

To address the limitations of traditional CTF challenges and existing training platforms, we developed SecGen, Hackerbot, and Hacktivity.

SecGen, our open-source Automatic Problem Generation (APG) framework, generates a wide variety of security challenges. Compared to previous narrowly-scoped APG systems, SecGen allows for the flexible specification of randomised configurations of networks of VMs, including servers, desktops, and attack VMs. This framework supports full-length security labs, practical tests, and realistic scenarios representing organisational networks for offensive and defensive assignments.

Hackerbot, an interactive chatbot, plays an important role in many of SecGen's lab-based scenarios. It guides students through interactive challenges, verifies completion, and outputs reward flags upon successful task completion, providing an engaging automated adversary for both defensive and investigative tasks.

Hacktivity builds on SecGen by providing a platform for managing and accessing VMs for cyber security education. It integrates randomised labs and challenges into security modules, offering a virtual learning environment (VLE) and a virtualisation management platform.

3.2 Hacktivity Architecture

Hacktivity is a Ruby on Rails application, which integrates closely with SecGen, running VMs on virtualisation platforms such as oVirt and Proxmox to deliver a comprehensive and dynamic learning environment for cyber security education. The Hacktivity engine runs via Passenger and Nginx, and leverages Redis for caching and background job processing and PostgreSQL for database management. The platform utilises web sockets to push updates from the server to clients in real-time.

3.3 Automated Provisioning of VMs via SecGen

Hacktivity bulk provisions batches of VM sets. Hacktivity creates VMs by adding jobs to SecGen's batch_secgen service to generate VM sets that make up a scenario – for example, a lab or CTF may involve a set of VMs with an attacker (such as Kali Linux) and vulnerable servers to attack, or desktop systems to manage.

Hacktivity uses SecGen to automatically provision these VMs onto the chosen virtualisation platform, currently either oVirt or Proxmox. This includes creating isolated networks and installing potentially vulnerable software and configurations as specified by the SecGen challenge scenarios. The SecGen framework is described further in Sect. 4.

3.4 Managing Pools of VM Sets

Once SecGen has generated the virtual machines (VMs), Hacktivity has responsibility over their management. Hacktivity manages pools of pre-generated VM sets to ensure generated VMs (with randomisation) are readily available when users need them. This pre-generation approach reduces wait times and ensures students can start their tasks on demand. The platform tracks and instantly allocates these VM sets to users or teams upon request.

Once a user has finished with a set of VMs, they can be restored to their original state using snapshots and can then be released back into the pool of available VM sets.

3.5 VM Management and Operations

Hacktivity integrates with multiple datacenter clusters, dynamically moving VMs in and out of clusters and between server nodes to maintain load balance across available server resources. This management is facilitated through virtualisation APIs such as Proxmox and oVirt. Live updates of node resources and VM states are achieved through polling of centralised server node logs and via event log APIs.

Users can instruct Hacktivity to start and stop VMs as needed. Other VM control actions users can take include snapshots and disk mounting.

Desktop access to VMs is facilitated via SPICE (Simple Protocol for Independent Computing Environments), using a user-client VirtViewer program. SPICE provides a high quality desktop experience with features including full HD resolution, copy-and-paste support, and file transfer.

3.6 Security and Authentication

Hacktivity itself takes a proactive approach to security. Role-based access control limits user permissions based on types of users. Staff access is scoped to specific organisations to limit cross-organizational data access. Hacktivity integrates with Azure Active Directory for secure student ID linking. Hacktivity enforces strong password policies to guard against brute-force attacks. Request rate-limiting mitigates against denial-of-service attempts and automated scanning. Multi-Factor Authentication (MFA) is required for all logins, which adds a layer of protection, reducing the risk of unauthorised access. Content Security Policy (CSP) is in place to provide an additional layer of defence against the potential for cross-site scripting (XSS). Recaptcha defends against automated bot attacks on forms and login pages. For an added level of security, MFA reprompts are required for some sensitive actions, ensuring that even if a session is compromised, critical operations still demand additional verification.

4 SecGen Framework

4.1 Design and Implementation

The SecGen framework is a Ruby application designed to generate randomised virtual machines (VMs) for security training scenarios, based on specified scenario constraints. The configuration language for SecGen is XML-based, allowing detailed descriptions of available vulnerabilities, services, networks, users, and content. Upon reading these configurations, SecGen applies logic to randomise the scenarios and leverages Puppet and Vagrant to provision the required VMs [11,14].

4.2 Modular Architecture

SecGen's modular architecture is central to its design, where VMs are created using a code-based approach using reusable and parameterised modules. The architecture is inspired by the Metasploit Framework, with a directory structure that promotes modularity and reusability. This structure includes directories for different types of modules such as vulnerabilities, services, utilities, encoders, and generators. For instance, vulnerability modules may specify the installation of insecure software, while service modules configure network services, and utility modules install tools and libraries on the VMs. Encoder and generator modules transform and generate data used in scenarios. Modules can include Puppet code for VM configuration or local Ruby code for data randomisation or transformation. The modular approach includes default inputs into parameters. For example, the default input for a username could be the result of a randomised username generator module, unless an input is specified by the scenario. Modules also define dependencies, and conflicts between modules, ensuring flexibility and reusability of modules for creating scenarios. The root of each module's directory contains a secgen_metadata.xml file, which defines the module's attributes: for example, in the case of a vulnerability module it can include details such as CVE information, privilege access levels, and descriptions of the vulnerability. This metadata allows for filtering and selecting appropriate modules based on scenario requirements [14].

4.3 Code-Based Specification of Scenarios

SecGen scenarios are specified using a code-based approach, which allows for detailed and flexible descriptions of the VMs to be created. Scenarios can define constraints and properties for the VMs, such as specifying certain types of vulnerabilities or services. For example, a scenario might specify the creation of a system with a remotely exploitable vulnerability leading to user-level compromise, and a locally exploitable flaw for root-level compromise, in which case the actual vulnerabilities employed would be selected from a large range of software, which would be further randomised in its configuration. Alternatively, a scenario can be more specific, and dictate exactly which software to install and how to configure it (by specifying values for any parameters).

The XML configuration language enables these detailed specifications, and the framework's logic applies randomisation to create unique instances of the described scenarios. Figure 1 illustrates an example snippet of a scenario.

Through this combination of modular architecture and code-based scenario specification, and randomisation, SecGen provides a powerful tool for generating meaningful and varied security challenges.

5 Integration with CyBOK

5.1 CyBOK Mapping and Curriculum Alignment

Hacktivity and our underlying SecGen framework integrate extensively with CyBOK, the Cyber Security Body of Knowledge. CyBOK provides a detailed

```
<system>
  <system_name>linux_server</system_name>
  <base distro="Debian 12"/>
  <vulnerability
    module_path=".*/unrealirc_3281_backdoor">
    <input into="strings_to_leak">
      <generator type="flag_generator" />
    </input>
  </vulnerability>
  <vulnerability
    privilege="root_rwx"
    access="local">
    <input into="strings_to_leak">
      <generator type="flag_generator" />
    </input>
  </vulnerability>
...
</system>
```

Fig. 1. Snippet of a SecGen scenario (with a specific remote vulnerability, a random privilege escalation vulnerability, and randomised flags).

overview of the core concepts and techniques within the field of cyber security, encompassing 5 Categories, a wide range of 21 Knowledge Areas (KAs), each composed of Topics, and content. For example, the category "systems security" currently has 5 KAs including "Authentication, Authorisation, and Accountability", which has topics including "authorisation", covering concepts such as "access control" and "access control lists (acls)". Figure 2 illustrates the relationship between categories and KAs.

CyBOK can be used by educators to design, update, and align educational content within cyber security domains. In the UK, CyBOK mapping is an important aspect of National Cyber Security Centre (NCSC) degree accreditation. CyBOK can also be a useful tool for learners to reflect on their own personal development.

5.2 SecGen CyBOK Mapping of Content

SecGen/Hacktivity boasts a robust curriculum currently comprising 118 labs, 40 multi-step Capture The Flag (CTF) challenges, and 71 lectures that have been meticulously mapped to CyBOK. These span a diverse range of technical topics aligned with CyBOK's KAs, Topics, and content.

SecGen's XML schema includes mapping representation of CyBOK metadata. This provides a structured format for representing how each lab, lecture, and challenge maps to specific CyBOK knowledge. The use of XML ensures consistency and clarity in documenting these mappings (see Fig. 3 for an example of the XML used for CyBOK mapping).

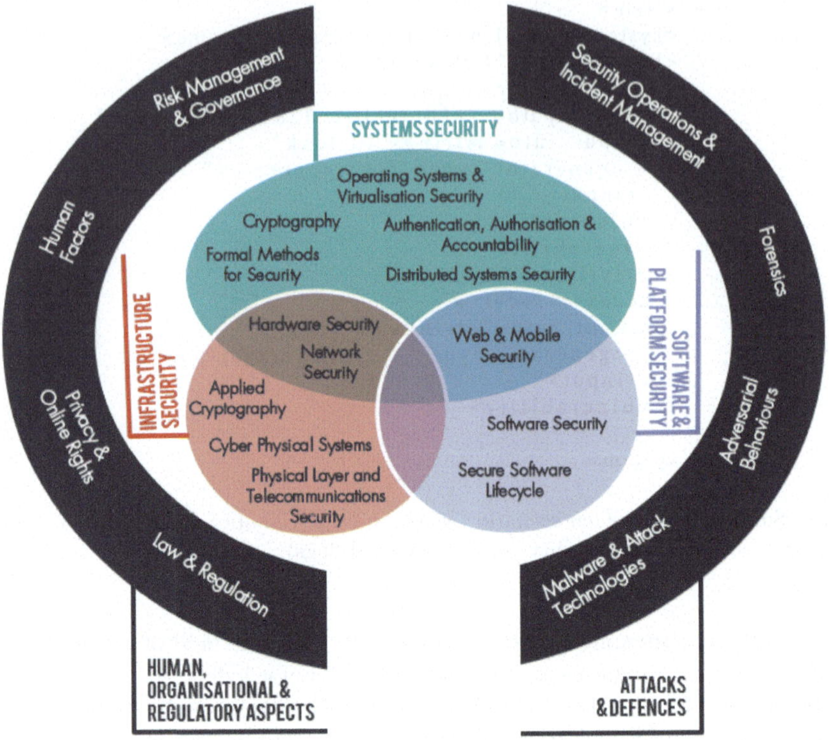

Fig. 2. CyBOK Categories and Knowledge Areas (KAs) [10].

This detailed mapping has enabled us to produce indexes of the mapped lab scenarios, CTF scenarios, and lectures.[1]

5.3 Hacktivity CyBOK Mapping of Content

CyBOK mapping within Hacktivity is visible across multiple levels of the learning environment. Each lab challenge (based on a SecGen scenario) and for any embedded lecture videos, CyBOK mappings are visible to end users, providing learners with clear insights into how each exercise contributes to their understanding and application of CyBOK principles. In addition to showing these individually, these are combined to provide a detailed view of what is covered.

Furthermore, summarised CyBOK overviews for entire modules are presented as high-level spider/radar graphs. This allows users to see the detailed alignment of Hacktivity's content with CyBOK at various granularities, aiding both learners and educators in understanding the educational objectives being addressed. Figure 4 provides mappings of modules to CyBOK.

[1] https://github.com/cliffe/SecGen/blob/master/README-CyBOK-Scenarios-Indexed.md.

```
<CyBOK KA="AAA" topic="Authorisation">
  <keyword>access control</keyword>
  <keyword>enforcing access control</keyword>
  <keyword>ACCESS CONTROL – DAC (DISCRETIONARY ACCESS CONTROL)</keyword>
  <keyword>Vulnerabilities and attacks on access control misconfigurations</keyword>
</CyBOK>
<CyBOK KA="OSV" topic="Primitives for Isolation and Mediation">
  <keyword>Access controls and operating systems</keyword>
  <keyword>Linux security model</keyword>
  <keyword>Unix File Permissions</keyword>
  <keyword>Filesystems, inodes, and commands</keyword>
  <keyword>umask</keyword>
</CyBOK>
<CyBOK KA="OSV" topic="Role of Operating Systems">
  <keyword>mediation</keyword>
</CyBOK>
```

Fig. 3. CyBOK metadata within a SecGen scenario.

Furthermore, Hacktivity tracks, on an individual basis, each user's experience with CyBOK, which is explained further in Sect. 6.

CyBOK provides a shared vocabulary for describing cyber security knowledge, which can inform course design and clearly communicate the focus areas of a degree program. Hacktivity demonstrates that CyBOK can be applied in practical training environments, to assist learners to comprehend the concepts they are studying and applying.

6 UX, Gamification and User Engagement

6.1 User Access and Interaction

Users sign up for events and modules through the Hacktivity platform. Registered users, including students from registered institutions, users from registered organisations, and guests, can browse available events and sign up for those they are eligible to join. For instance, modules/courses they are enrolled on, CTF hacking challenges, such as hack boxes, and special events during open days.

Accessing labs and challenges is straightforward. Users can assign themselves a set of VMs, which is immediately allocated to them from the pre-built pool of VM sets for the challenge. Each VM set is unique, typically with randomisation such as different usernames, passwords, and IP addresses, VM configurations, selection of vulnerabilities or challenges, and configuration of challenge details (amongst others), to ensure a personalised learning experience, while improving replayability and mitigating cheating and unfair practice.

Users can activate and deactivate challenges as needed. Activating a challenge deploys the VMs into the virtualisation infrastructure (as described in Sect. 3), ensuring users have access to the resources required for their current tasks (Fig. 5).

6.2 VM Control

Hacktivity enables users to control virtual machines (VMs) with ease. An intuitive interface enables starting, stopping or shutting down, and taking or rolling back snapshots.

Once a VM has started, users can interact with it through a graphical interface or via network access (via another VM in the scenario), depending on the specifics of the challenge. Hacktivity generates .vv files to enable SPICE access via a remote-viewer application.

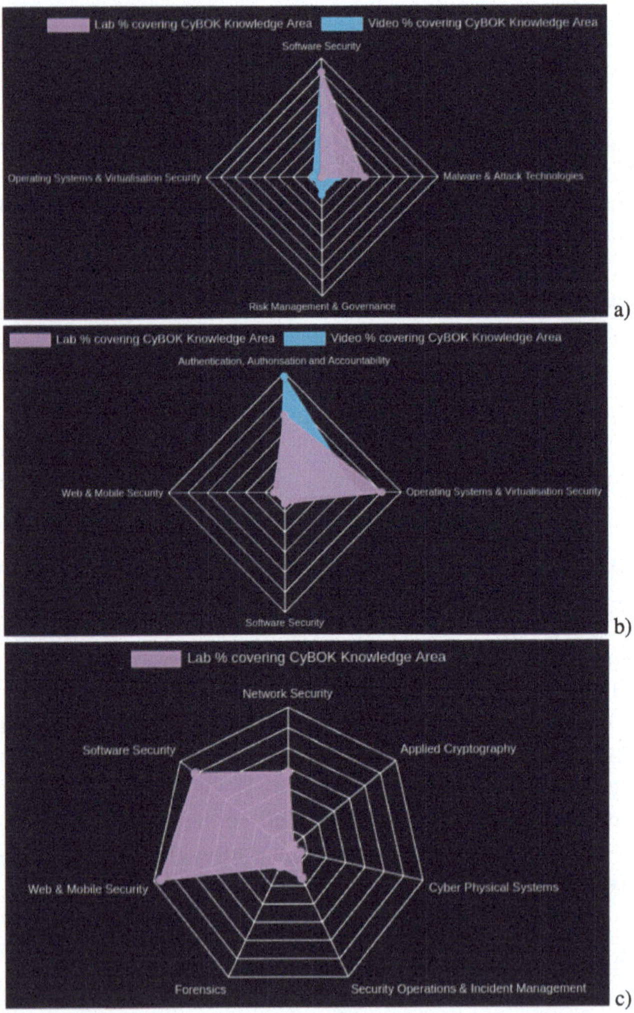

Fig. 4. CyBOK mapping for modules: a) Software Security and Exploitation, b) Systems Security, c) Web and Network Security, d) Ethical Hacking and Penetration Testing, and e) Incident Response and Investigation.

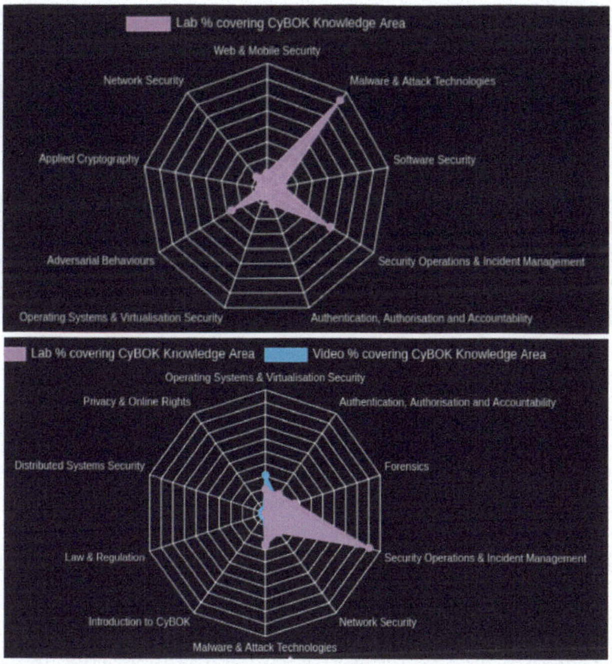

d)

e)

Fig. 4. (*continued*)

6.3 Flags and Scoring

Hacktivity incorporates a flags-based scoring system, providing instant feedback and automated marking for completed challenges. When users solve challenges, they receive flags, typically formatted as "flag{something_random}". These flags are submitted through the Hacktivity platform for immediate scoring.

Each challenge (topic/lab) typically involves a number of flags to solve, and can provide a weighted contribution towards an overall module/course score. For example, if the first lab in a course has 10 CTF flags, and that topic contributes 3%, then completing 5 of those flags means that lab has a score of 50% and contributes 1.5% towards the course mark. Submitting a flag immediately updates the users scores and progress.

Hacktivity also includes education-focussed scoring features, such as timed CTF tests, and late penalties.

Hacktivity is designed to give users real-time feedback on their achievements, enhancing the learning process by allowing them to track their progress continuously.

6.4 Scoreboards and Weekly Streaks

Hacktivity's progress monitoring features enable users to track their progress and achievements in real-time. Hacktivity uses scoreboards, stats, and weekly streaks to gamify the learning experience further.

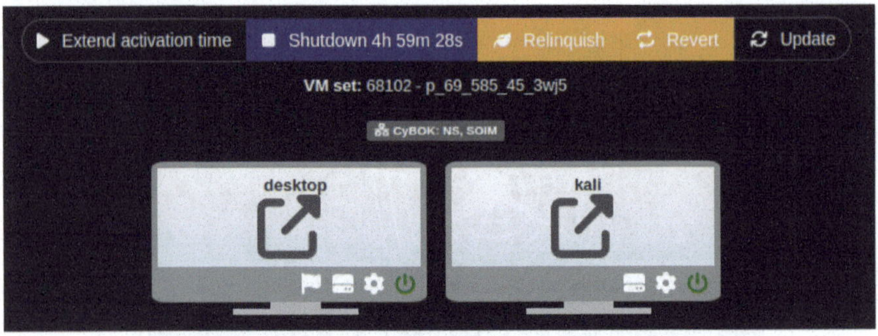

Fig. 5. VM set controls in Hacktivity.

Scoreboards rank users for each separate lab challenge or overall competitive event, fostering a sense of competition and encouraging continuous engagement. Furthermore the first and the most recent user to complete each challenge are celebrated. Whenever a user submits a flag, all other users are notified, encouraging engagement on the platform.

Hacktivity also gathers detailed stats on each user, including their engagement with the platform as a whole and with each course, this includes number of flags solved, days active, and detailed results.

Weekly streaks are another motivational tool, rewarding users for consistent participation and achievement over time. So long as the user completes a flag each week, they continue their weekly streak, with a record of the number of solved flags and challenge completions, and length of current and longest streak.

6.5 Personalised CyBOK Experience

Hacktivity tracks the learning that each user has engaged in, and leverages the extensive CyBOK mapping of content to present this information to users. Hacktivity presents each user with visualisations of their learning journey in terms of CyBOK topics, topics and content, which is designed to help them reflect on what they have learned and knowledge they have put into practice.

6.6 Knowledge Experience Points (KXP)

Hacktivity calculates a "Knowledge Experience Points (KXP)" score for each of the CyBOK KAs, based on the labs and flags that have been solved and completed and based on other learning materials, such as videos. This approach incorporates gamification, incentivising continuous learning, but also importantly producing useful information that helps the user (and their supervisor/tutor) to reflect on their strengths and the focus of their learning and achievements.

Figure 6 demonstrates this by showing the KXP visualisations for three different students currently using Hacktivity. Each of the three students have distinct

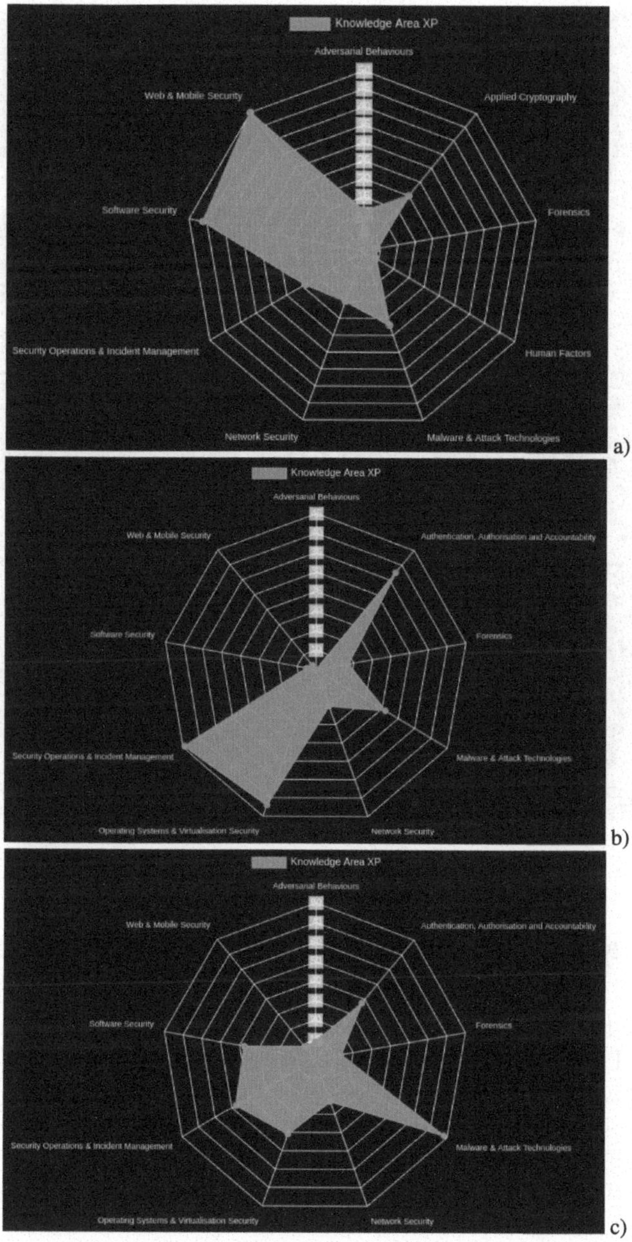

Fig. 6. KXP visualisations for a (a) second, (b) third, and (c) third year university student.

personal CyBOK profiles representing the challenges and topics that they have engaged in, and identifies strengths.

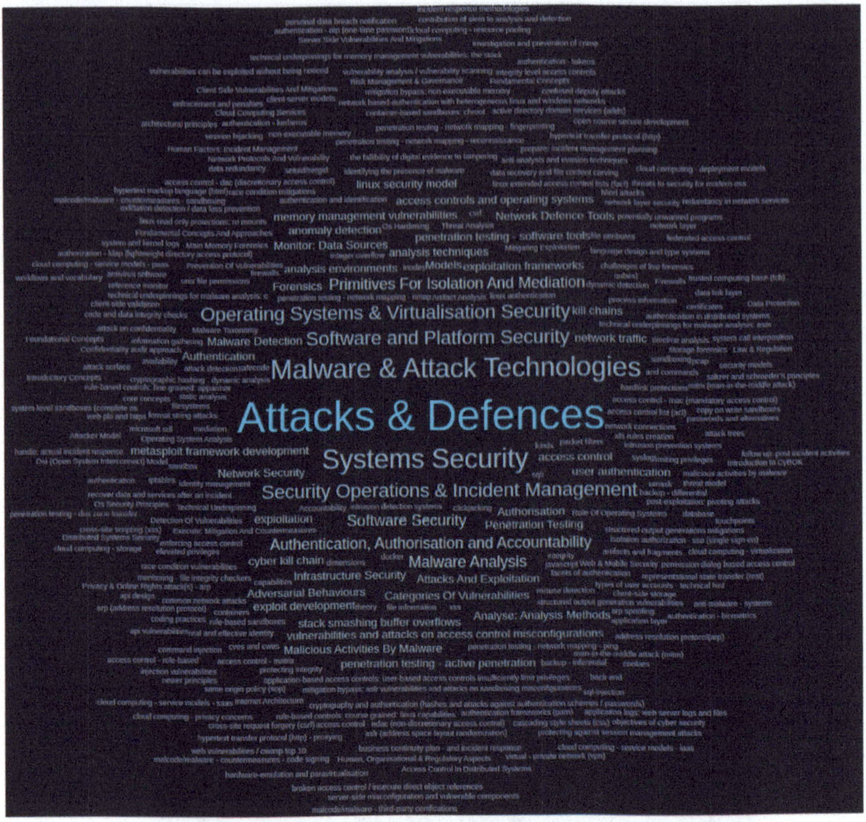

Fig. 7. CyBOK word cloud visualisation of a user's experience.

6.7 CyBOK Word Clouds

Furthermore, Hacktivity generates word clouds to visually represent the concepts and skills that users have encountered throughout their learning journey. As illustrated in Fig. 7, which is from a real user, these word clouds provide a reflective summary.

By integrating with CyBOK, Hacktivity not only aligns its educational content, clearly states what knowledge is covered in each challenge and course, but also enhances the learning experience through personalised tracking and gamified elements.

7 Implementation and Case Studies

7.1 Integration Into Academic Courses and NCSC Accredited Degrees

At Leeds Beckett University (LBU), Hacktivity is a cornerstone of our NCSC provisionally accredited cyber security degrees, with approximately 300 students on modules involving Hacktivity. All of our cyber security modules incorporate hands-on labs and marks from Capture the Flag (CTF) challenges.

We have had success integrating Hacktivity throughout our entire suite of cyber security modules. Module titles include:

- Ethical Hacking and Penetration Testing
- Cyber Security Landscapes
- Web and Network Security
- Systems Security
- Incident Response and Investigation
- Reverse Engineering and Malware Analysis
- Software Security and Exploit Development

Most of these modules have both distance learning and local cohorts.

Each module includes lectures. In some cases these are prerecorded and directly embedded within Hacktivity, in other cases they are delivered in person or live online.

Learning tasks, such as labs and assignments, for these modules are delivered via Hacktivity. Students claim VMs and work through lab instructions, which are embedded within Hacktivity.

For all of these modules 100% of the technical digital learning tasks that students complete are completed directly via Hacktivity. The weekly labs involve security challenges for students to solve (randomised per student), which contribute to marks. Assessment uses include practical CTF-style timed tests, randomised systems for teams of students to perform security audits of, and VMs for students to configure security according to requirement specifications. Non technical topics and tasks, such as risk management, and any written aspects of assignments, and some (non-CTF) tests, are completed off-platform.

All these modules include CTF challenges that contribute to students' marks. Technical concepts covered by randomised challenges include stages of attack and penetration testing (including infrastructure scanning, exploitation, post-exploitation and pivoting, topics within web security and OWASP top 10, such as XSS, SQLi, CSRF), malware and reverse engineering, software vulnerabilities, authentication, access controls, and system hardening (including ACLs, setuid, MAC), networking and attacks, containers and sandboxing, monitoring integrity, IDS (detecting real automated attacks over the network), exfiltration detection, live and dead system analysis for incident response, phishing, applied cryptography, secure software development and bug hunting, fuzzing, exploit development (including buffer overflows, and against mitigations such as ASLR, NX), and many other topics.

Additionally, LBU has a Team Project module which uses Hacktivity for a randomly generated penetration testing target allocated to teams.

We recently completed CyberASAP (Cyber Security Academic Startup Accelerator Programme) and have started commercialisation of the platform. We have signed up our first external academic customers, and are helping them incorporate Hacktivity into a new course they are running next academic year.

Detailed mapping of content to CyBOK helps ease the process, and the content can be adopted for use in a variety of cyber security education programmes.

7.2 Effectiveness in Teaching

Hacktivity has proven effective in enhancing the teaching of cyber security concepts. Hacktivity reduces grading burden through automatic marking; provides a safe lab environment for real offensive security tasks; provides a full desktop environment to students and staff; removes the need for most academic staff to be involved in virtualisation management; provides a VLE with access to learning resources and marks downloads and student engagement; mitigation against collusion and cheating; and encourages and develops student adaptability.

All LBU cyber security modules now include elements of automatically marked work, significantly reducing the grading burden on academic staff. Previous evaluations of our initial integration of a gamified approach found that gamification increases student enjoyment, though our initial implementations required significant marking overheads [12]. It was clear that without automated systems in place this approach was not sustainable. Our Hackerbot and delivery with Hacktivity have now 100% automated the marking of the weekly learning tasks from that module. While this doesn't remove the need for other forms of assessment, automated CTF challenges provide the benefit of regular assessment and feedback, without burdening staff.

Hacktivity successfully provides a safe lab environment with all the software tools and challenges to teach an entire curriculum of content, for all the technical cyber security modules in BSc, MSc, and MEng degrees. Isolated networking of VMs means that students are given a safe environment for experimenting with and learning from offensive techniques and malware, performing real attacks and security tasks.

A full desktop experience is provided, without the need for students to maintain their own attack VMs or VPNs.

Remote access has proven crucial for supporting distance learning students, and means that any registered users can access the labs from home. During the COVID-19 pandemic Hacktivity facilitated a straightforward transitioning of local students to remote delivery.

Academic staff have access to students' VMs, via the same full desktop experience, with additional access to all root accounts and servers within a student's VM sets. This level of access is not required by students in order to complete the challenges, but it enables teachers to support students, by inspecting or exploring student's VMs directly. This has proved particularly helpful when remotely supporting distance learning students.

Provisioning VMs is taken care of by Hacktivity, and maintained by admin staff, which has meant that most academic staff do not need to be involved in any virtualisation management. The implementation details, such as clusters, nodes, VMs, cloning of templates, and so on, are hidden from most users of the platform, including academic staff. This saves staff time and effort.

Hacktivity provides a virtual learning environment (VLE) with learning resources such as videos and lab instructions, organised around learning modules. Hacktivity provides a CSV download of scores, which can easily be used to populate marks on an institution's VLE, such as Blackboard or Moodle. Timely progress tracking and statistics on user engagement also empowers staff to support students.

Randomisation provides mitigations against collusion and cheating. The randomisation of the details of challenges means students cannot simply copy solutions or flags from each other.

The randomisation also ensures that students develop adaptability, rather than relying on rote learning. For example, during a timed CTF test, a student who attempted to memorise and execute commands without understanding them failed because the details such as usernames and IP addresses had been randomised, confounding the commands that they had attempted to memorise without understanding.

During a BCS (The Chartered Institute for IT) degree accreditation visit, the Accreditation Visit Panel identified "Hacktivity and related learning and teaching approaches" as an example of best practice in the sector.

7.3 Student Experience and Reception

Student feedback has been positive, as evidenced by high NSS scores and frequent praise in module evaluations. Students appreciate the practical, hands-on approach to learning that Hacktivity provides.

We have found that this assessment approach with frequent CTF challenges fosters intrinsic and extrinsic motivation, which encourages students to engage regularly and deeply with the material. Anecdotally, compared to modules and labs where we do not assess lab outputs, we see a much greater completion rate. Furthermore the flag-based approach is enjoyed by students, who enjoy completing the CTF challenges, and the feedback that's provided.

The use of Hackerbot has provided an automated adversary for students to engage interactively with, and has automated tasks that otherwise wouldn't work without working with other students (which is not always possible), and making the learning process more immersive and enjoyable [13].

Hacktivity also enables users to engage in extra-curricular hacking challenges, in the form of "Hack boxes". We provide scenarios with multiple step CTFs. Some students have really enjoyed engaging in these extra unguided challenges.

Feedback from students has included:

- "Hacktivity is a powerhouse; I often forget how seamless the VM access is."
- "I enjoyed the Hacktivity weekly challenges a lot."
- "This course was informative and rewarding due to the CTF challenges. [...] I like how the module, through the use of flags, urges you to apply the theory from the lectures and the labs into practical use."
- "The use of the Hacktivity platform significantly enhanced the learning experience by providing a hands-on, interactive environment that made complex theories more accessible and engaging."

- "Very fun and interactive with satisfying challenges."

Anonymous student feedback

7.4 Benefits of Open Source Framework

SecGen, our open source backend framework, is open source software published on GitHub, with 2,592 stars (likes) and 314 forks. While GitHub does not report on download totals, the last 14 days (as of writing) involved 1,139 Clones by 32 Unique cloners.

We have heard from various universities making use of SecGen, including from staff/students at other universities in the USA, UK, Vienna, and Hong Kong.

Because the underlying SecGen project is open source, staff and students can be involved in improving and developing new content. As captured by GitHub 26 separate developers have contributed. This includes staff and students from Leeds Beckett University, University of Birmingham, USA Fort Hays State University, and a number of other online contributors.

Because the approach taken by SecGen is code-based and modular, when new content is created then existing scenarios can automatically make use of these to add more variety. For example, this can include new vulnerable software, and ways of encoding or generating data.

We deployed a survey to users of SecGen: 20 separate respondents reported using SecGen with groups of university students, 15 with industry professionals, 11 with hobbyists, 3 with secondary/high school students, and 1 person used SecGen with primary school students. SecGen has been reportedly used to provide lab environments/lab sheets; host capture the flag competitions; provide systems for practising specific attacks; perform blind security audits, and for learning from the source code. The number of users ranged from 1 user through to 5000 participants.

8 Conclusion

Hacktivity Cyber Security Labs offers a comprehensive solution to challenges faced in traditional cyber security education. By leveraging randomisation, automated virtual machine (VM) provisioning, and integration with the Cyber Security Body of Knowledge (CyBOK), Hacktivity provides an immersive, dynamic, and scalable learning environment. The platform's ability to generate unique challenges reduces the risk of academic dishonesty and ensures a varied learning experience, better preparing students for real-world cyber security scenarios.

Furthermore, Hacktivity's extensive use of automated chatbots and interactive challenges enhances student engagement and provides instant feedback, making learning more interactive and effective.

The close integration with the Cyber Security Body of Knowledge (CyBOK) further distinguishes Hacktivity by ensuring that educational content is mapped to the specific knowledge that is covered. This is crucial for curriculum development and accreditation, and offers educators a structured framework to adopt Hacktivity content into existing courses, and design and update courses. For students, CyBOK mapping is used to provide methods for tracking their learning progress, additional elements of gamification, and assist them to reflect on all the concepts that they have learned and applied.

SecGen is a free and open source resource that is available to the community, and Hacktivity is a hosted platform.

References

1. Burket, J., Chapman, P., Becker, T., Ganas, C., Brumley, D.: Automatic problem generation for capture-the-flag competitions. In: 2015 USENIX Summit on Gaming, Games, and Gamification in Security Education (3GSE 15). USENIX Association (2015)
2. Chapman, P., Burket, J., Brumley, D.: PicoCTF: A Game-Based Computer Security Competition for High School Students. In: 3GSE, 2014 (2014). Retrieved from https://www.usenix.org/system/files/conference/3gse14/3gse14-chapman.pdf
3. Chung, K.: Lowering the Barriers to Capture The Flag Administration and Participation. In: 3rd Annual Symposium on Advances in Security Education (ASE 17) (2017). Retrieved from https://www.usenix.org/system/files/conference/ase17/ase17_paper_chung.pdf
4. CSAW. CSAW Capture the Flag (2024). Retrieved from https://www.csaw.io/ctf
5. Facebook Inc. FBCTF - Meta Archive (2024). Retrieved from https://github.com/facebookarchive/fbctf
6. Feng, W.: A Scaffolded, metamorphic CTF for reverse engineering. In: 2015 USENIX Summit on Gaming, Games, and Gamification in Security Education (3GSE 15) (2015). USENIX Association. Retrieved from https://www.usenix.org/system/files/conference/3gse15/3gse15-feng.pdf
7. Hack The Box. Hack The Box (2024). Retrieved from https://www.hackthebox.com
8. Hulin, P., et al.: AutoCTF: Creating Diverse Pwnables via Automated Bug Injection. In: 11th USENIX Workshop on Offensive Technologies (WOOT 17) (2017). USENIX Association
9. Immersive Labs. Immersive Labs (2024). Retrieved from https://www.immersivelabs.com/
10. Rashid, A., Chivers, H., Danezis, G., Lupu, E., Martin, A., Schneider, S.: The Cyber Security Body of Knowledge (2021)
11. Schreuders, Z.C., Ardern, L.: Generating Randomised Virtualised Scenarios for Ethical Hacking and Computer Security Education: SecGen Implementation and Deployment (2015). Retrieved from http://eprints.leedsbeckett.ac.uk/1479/
12. Schreuders, Z.C., Butterfield, E.M.: Gamification for Teaching and Learning Computer Security in Higher Education. In: 2016 USENIX Workshop on Advances in Security Education (ASE 16). USENIX Association (2016). Retrieved from https://www.usenix.org/system/files/conference/ase16/ase16-paper-schreuders.pdf
13. Schreuders, Z.C., Shaw, T., Muireadhaigh, A.M., Staniforth, P.: Hackerbot: Attacker Chatbots for Randomised and Interactive Security Labs, Using SecGen and oVirt (2018)

14. Schreuders, Z.C., Shaw, T., Shan-A-Khuda, M., Ravichandran, G., Keighley, J., Ordean, M.: Security Scenario Generator (SecGen): A Framework for Generating Randomly Vulnerable Rich-Scenario VMs for Learning Computer Security and Hosting CTF Events (2017)
15. TryHackMe. TryHackMe (2024). Retrieved from https://tryhackme.com

Teaching Adversarial Thinking by Having Students Circumvent Exam Rules

Matthew Bowden, Tom Chothia, Anna Clee, Sam Collins(✉), Jacqueline Henes, and David Oswald

University of Birmingham, Birmingham, UK
{mxb1143,axc1017,sxc1327,jkh703}@student.bham.ac.uk, {t.chothia,d.f.oswald}@bham.ac.uk

Abstract. This paper describes a live exercise in adversarial thinking, deployed as part of an advanced undergraduate module: *Security of Real World Systems*. Delivered as a mock exam, students must answer a single question given two weeks in advance: write the first sixty digits of π. We present a student attacker model, from which we construct a set of exam rules, all containing deliberate oversights. Students must analyse the rules for vulnerabilities, develop a solution, and successfully deploy it in a mock exam invigilated by the module team. We describe how we handled communication with students, invigilation, and marking such an exercise. We run an educational evaluation, categorise student solutions, and gauge student opinions on the exercises from feedback on the module, as well as a direct focus group. Finally, we discuss some more salient lessons from the exercise and how we might address them as we take our work forward.

1 Introduction

Teaching the practical skills required in the field of cybersecurity is innately difficult - the field does not lend itself well to pure bookwork, and exercises can often be lengthy in their setup and difficult in their deployment. Additionally, a balance must be found between theory, practical skills, and fostering the creative approaches that the field demands in students.

In this work, we describe an exercise in adversarial thinking that we ran as part of the 3rd year B.Sc. course *Security of Real-World Systems* at the University of Birmingham, UK. This is an advanced cybersecurity module that covers topics such as buffer overflows, ROP attacks, side-channel analysis, and protocol attacks. A recurring theme - and key learning objective - of the course is that understanding the *rules* of a system well often makes it possible to discover exploits that were not foreseen when it was created. For example, understanding the layout of the stack and calling convention used in x86_64 allows for return-to-libc and ROP attacks, and understanding the instruction-dependent timing and power usage of CPUs reveals a side-channel attack on AES.

Based on an idea by Conti and Caroland [3], we set a mid-term "exam" that asked students to write out the first sixty digits of π. This exam-like assignment

P. Legg et al. (Eds.): CSE 2024, LNNS 1213, pp. 75–95, 2024.
https://doi.org/10.1007/978-3-031-77524-6_5

had a carefully crafted set of rules that would be enforced as written. While initially seeming impossible to solve, the exam rules contained enough loopholes for the student to complete the task. Conti and Caroland encourage their students to cheat and break the rules [3]; conversely the objective of our exercise is for the students to analyse the rules; find weaknesses in them; and exploit them; this is in line with the key learning objectives of the course.

We ran the mid-term exam by replicating the format of our institution's exam procedure, with the exercise worth a token 2% of the overall course mark. The students demonstrated a wide range of creative solutions with a mix of approaches both expected and unexpected. Student feedback on the test indicated that this was very popular, and that the learning objective was well understood.

In summary, the contributions of this work are:

- A student attacker model designed to promote outside-the-box analysis and adversarial thinking.
- A framework for developing mock exam rules with exploitable oversight, as well as our ruleset built to match the student attacker model.
- Guidance on communication with students before such an exercise, invigilation, and assessment .
- An educational evaluation of our results.
- Details of practical lessons learned, with guidance for other institutions who may wish to deploy a similar exercise.

Our ruleset can be found in Appendix and further details can be found at https://www.cs.bham.ac.uk/~tpc/ExamExercise/.

2 Background

2.1 Live Security Exercises

Cybersecurity lies apart from other disciplines in the fields of science and engineering due to its heavily adversarial nature. Progress in cybersecurity is not just driven by core count and computational complexity, but by new methods to obfuscate, outmaneuver, and trick adversaries. Therefore, live security exercises and gamified challenges have become popular in the field [4], chief among them being Capture The Flag (CTF) exercises. Past research [2] has neatly classified CTFs into the following two categories:

- **Adversarial (Attack/Defence) -** whereby participants/teams must protect their vulnerable services while exploiting the other teams vulnerable services.
- **Jeopardy Style -** where there are a wider range of standalone attack-only challenges which doesn't require/support interaction between participants.[1]

[1] Davis et al. [4] suggests that if you want a defence-only CTF, you need only host something online and wait for it to be attacked.

This distinction is important, as there has been further research into the benefits and respective drawbacks of each style of CTF. Davis et al. [4] suggests that students see the defence side of adversarial CTFs to be more of an annoyance, suggesting a mirror to real life sports where the star athletes on a team tend to be the ones who score points. This has given rise to the Jeopardy-style CTF, which are now far more common, especially for learning and assessment at a more beginner to intermediate level.

Other forms of live security exercise exist and have been successfully deployed both in and out of education. Many share common traits with more classic CTF challenges; we discuss this further in the following section (sect 3).

3 Related Work

Within the specific context of education, research shows the value of gamified live security exercises. The United States Air Force Academy (USAFA) found increased motivation in their undergraduate students when completing an attack-only CTF challenge [1] deployed in stages through the course, when contrasted against their regular assignments. Students were found to spend additional time researching complex and novel solutions to more difficult challenges, matched with an increase in enthusiasm.

A similar approach has been demonstrated in an alternate educational setting at UC Santa Barbara by Vigna [7] in teaching network security. Three challenges were deployed on the course, similarly to Carlisle, Chiaramonte, and Caswell [1], but as standalone exercises rather than a continuously evolving CTF. Importantly they find that more disruptive attacks, in their case Denial of Service on a network, interfere with the learning objectives of the exercise for other students. We take steps to avoid disruptive interference in our exercise, discussed in Sect. 4.3.

Ensafi, Jacobi, and Crandall [5,6] deployed their live security exercise as a variant of the social deduction game Werewolf on a mix of Undergraduate and Postgraduate students. Here students connect to the game server via SSH and are expected to exploit vulnerabilities in the server to deduce the identities of their classmates. They found mixed results; students exceeded expectations regarding the sophistication of attacks and demonstrated a high level of enthusiasm for the exercise. However, they struggled to make the game fair, with the students assigned the werewolf role having little chance to succeed when faced with more decisive tactics.

Making students cheat in a simulated exam has also been explored by US Cyber Command [3]. Deployed on a small group of students to encourage the development of a more 'devious' and adversarial approach to problem-solving, students were allowed complete freedom in how they cheated and were encouraged to directly break the rules given. This approach differs slightly from our own, where we allow cheating, but encourage bypassing the rules rather than breaking them; this encourages students to perform an analysis of the system.

4 Running the Exam

4.1 Attacker Model

When designing the exam, we first define a student attacker model as a framework to consider when constructing the rules. First, students have full knowledge of system rules as well as the system's environment from when the exam is announced. They are then able to use the allotted preparation time to analyse the system, consider vulnerabilities, and develop an attack. The objective is also announced ahead of the exam, but is of less consequence than the rules or environment. A key property of this model is that it is as liberal and open ended as possible within the bounds of safety, the law, and disruption to other students. Students may attempt any attack, including social engineering and insider attacks[2]. We only had one attempted phishing attempt before the exam, shown in Fig. 1.

Worth considering is the factors that students do not have complete control over; for example, the environment the exercise takes place in. Telling the students of the environment (e.g. the exact room number in a building) in advance gives information that can be considered as part of their design, however relevant setup and precautions should be taken so that the exercise difficulty is not undermined. In our case, many suspicious items were removed from the examination room shortly before the start of the exam, as was said would be done in the rules. In one case, a student intended to use the computer at the lectern to retrieve the digits of π; in actuality this computer was unavailable for use at the time and they were unable to use this solution.

4.2 Exam Rules

The purpose of this exam was not to directly assess students understanding of specific course content, but to encourage them to engage with the target system as an adversary might. The exam rules should therefore support the abstract problem- solving abilities that cybersecurity work can demand, and in particular encourage novel solutions. To match the desired student attacker model, we released the set of exam rules and venue information two weeks before the exam, providing time for analysis and preparation. Alongside the rules we also released the objective of the exam, a single question: to list the first sixty digits of π.

Given the standard expectations of exams at the University of Birmingham[3], we believed students may need reassurance regarding the behaviour we expected of them as well as the consequences (or lack thereof) if they were

[2] We decided that attempting to convince an invigilator to render assistance is permissible, but we agreed that invigilator responses to such attempts should be pre-agreed, uniform, and fair.

[3] Please refer to https://intranet.birmingham.ac.uk/as/registry/exams/rules/rules-and-regulations.aspx for standard exam rules.

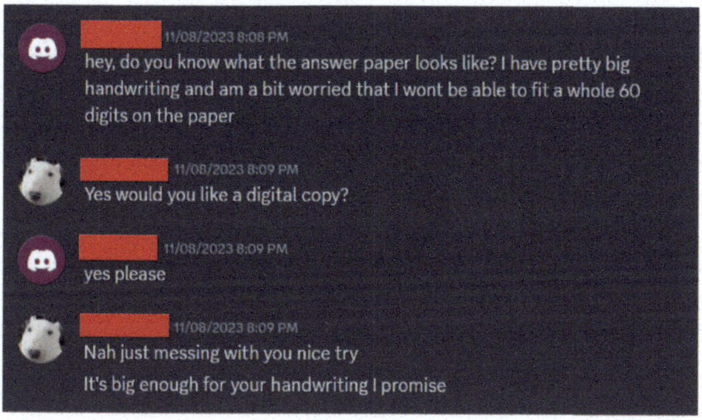

Fig. 1. The only social engineering attack we experienced, attempted a few days before the exam via Discord with a TA.

caught "cheating". Released as a part of the rule-set was a disclaimer paragraph reassuring students that circumventing the rules was encouraged.

> The rules of this "exam" will be enforced strictly and to the letter. There are no other rules. All the normal expectations of good, honourable behaviour that are required for the other exercises on this module, and other modules, are not required here. The normal description of academic misconduct and examination irregularities in the student handbook do not apply. If the rules outlined above are deemed to be broken, you will only fail this test, it will not be treated as a genuine case of an examination irregularity.

We encourage students not to cheat by breaking rules, but rather to circumvent them. We therefore disincentivise breaking the rules by adding the risk of failing the test if caught doing so. There is however a balance between there being a consequence of being caught and the severity of the consequences students are afraid of. We want them to have the confidence to attempt circumventing the rules, so we make sure they understand this will in no circumstance be treated seriously, beyond losing the marks of this exam, and there will be no disciplinary actions if caught. This is a complicated line to balance: students will need explicit reassurances as it goes against everything they have experienced so far in higher education. We discuss the importance of good communication with students and how we went about achieving this in more depth in Sect. 4.3.

We do not believe that any specific rules are necessary to run this exercise successfully; instead, rules should be designed in a way that invites students to find vulnerabilities in the system which they can the exploit. To this end we followed a set of design choices when constructing our rules, with some rules having one of these characteristics and others multiple:

- **Deliberately specific wording.** Using language which, in its specificity, leaves vulnerabilities, often by omission. Since only the exact

rules are enforced, anything not explicitly banned is implicitly allowed. ***Example:*** "Once the exam has started, students in the exam room must not talk to anyone else in the exam room, except the invigilators." - we both specify and repeat "in the exam room"; in this case students would be allowed to talk to someone outside the exam room including asking them to read out the answer while they write it down.
- **Deliberately vague wording.** Using language which is ambiguous in its meaning, or poorly defined in general. This gives students the opportunity to use the edge cases of a definition to their advantage. ***Example:*** "Students may not pass anything to each other during the exam." - Pass is poorly defined here; students would be allowed to show each other notes, or leave them for another to pick up later as long as they are not passed directly.
- **Omitted information.** Leaving out items of information from rules where they may be otherwise expected. If part of a rule is omitted then it is implicitly allowed by the system and students can therefore make use of it. ***Example:*** "No laptops, tablets, smartphones or smart watches. No calculators or any device that readily displays stored digits of pi. You must not use WiFi or Bluetooth wireless connections." - We've forbidden the most common consumer electronics. However by omission we allow other devices including desktops, servers, and those which support older communication standards such as radio or a wired connection.
- **Additional information.** Adding in information which is not required for the rule to make sense but will introduce an oversight for students to exploit. ***Example:*** "Students may choose their own seats. Once seated, they must stay in their seat until an invigilator gives them permission to leave. " - Here the addition of "once seated" leaves a loophole in the rule; if a student avoids sitting down, they may walk freely around the exam room.

The full ruleset is in Appendix.

Fig. 2. A device which vibrates the digits of π when prompted. The student kept this hidden but since it doesn't display the digits it doesn't break the rules.

These rules were designed specifically to be full of oversights that students can identify, as detailed above. Also, by specifying the exam's location long in advance, students can attempt to hide things or test the feasibility of their solutions. Invigilators were aware of some of the most obvious issues with the rules, and were broadly in agreement of what constituted a rule break in advance of the

Fig. 3. A device connected to the room's projector which emulates the standard university login screen, periodically replacing the time with the digits of π one at a time. This does break the rules but wasn't detected.

exam. Finally, it's important to leave some discretion to the invigilators; easily recognisable encodings such as sixty digits of π plus one should be disallowed as invigilators see fit.

4.3 Communication with Students in the Lead up to the Exam

4.3.1 Communicating Intent

Communicating the intent of this assignment is the difference between it being a successful exercise and enjoyable experience, instead of a stressful one. Students were informed of the exam two weeks in advance during a lecture and were able to ask clarifying questions at that time, but were also referred to the module discussion page to ask further questions. Separate to the rule-set we also emphasised to students some of the intentions behind the assignment in an announcement:

> There is not one single solution to this exercise. An issue with a lot of continuous assessment is that it points you to a single correct answer but real life problems aren't like that. There are a number of solutions we expect to see from students, but we also hope that some students will surprise us with solutions we haven't thought of.

4.3.2 Answering Questions

It is important that any answers given to students were available for all students to see so that no one had an unfair advantage. We encouraged them to reach out for clarifications if they were uncertain to avoid misunderstandings.

After the announcement a lot of attack ideas were poised directly to the module team, framed as "Would I be allowed to do ...?", to which the default response was "Please see the rules". We encourage a balance between helping students and letting them navigate the rules on their own. Answers should help students gain confidence in interpreting the rules independently and reassure them that the consequences of making mistakes will not be serious. Those running the exercise should not confirm or deny any ideas, as the aim of the exercise is for students to be analysing the rule set and interpreting them themselves.

One of the initial ideas suggested was to write the literal phrase "the first 60 digits of pi". This was a valid interpretation of our initial rules and so a valid answer, but this would have been too easy and reproducible, trivialising the exercise. Students would be able to avoid creatively coming up with varied solutions. Therefore, we announced and clarified to all students that this would not be a valid solution:

> Please note however, that writing the words "the first 60 digits of pi" on the exam paper will not be accepted, we need to see the actual numbers.

We would recommend to anyone doing a similar exercise themselves to add this exception explicitly in the rules.

4.3.3 Monitoring Student Apprehension

We believe that it is important to provide students with an outlet for their ideas and questions, allowing the team running the exercise to keep an eye on them. This way if the students are heading towards an interpretation which invigilators believe defeats the intent of the exercise, it may be caught and corrected. We reinforce that part of the uniqueness of this exercise is that students will come up with unexpected solutions, and that these should be encouraged whenever possible.

As we were performing this exercise for the first time, and wanted to keep an eye on the student thoughts and sentiments around this exercise, the teaching assistants also observed the students' Discord[4] channel for the module. We found the overall feeling was that of excitement, mixed with apprehension as to possible consequences. Two questions were particularly common:

1. What happens if we get caught with a banned item?
2. Could we get reported to the academic integrity committee?

Both these questions are addressed in the initial ruleset:

> Any student seen breaking the rules by an invigilator will have to leave the exam hall and will fail on the test...
> ...If the rules outlined above are deemed to be broken, you will only fail this test, it will not be treated as a genuine case of examination irregularity.

Additionally, we used a separate announcement to brief students on how the actual day would play out as this was a common a point of discussion:

> The invigilators will act as normal exam invigilators, we will not search you on the way in, but if we see something suspicious we will investigate. When you enter the room, we will instruct you to place bags at the back of the room or under your seat.

On the Discord server, which TAs volunteered to monitor to answer basic questions, some students suggested triggering the fire alarm, along with other

[4] Discord is an online chat and call platform. There is a popular unofficial Discord server aimed at students taking the Computer Science course at the University of Birmingham, often with sub-chats for specific modules.

disruptive ideas such as using large sound systems. While we believed these to be jokes, we reminded students that national laws still apply, general decent non-disruptive behaviour on campus is always necessary, and that they should be mindful of other classes in the building. Any solutions that could fatally disrupt the entire exam experience were explicitly ruled out.

Rather surprisingly, we had some students who were unsure about whether "the first 60 digits of pi" meant the first 60 digits after the decimal point. We clarified this to the students for the sake of complete transparency. It is an option to write the desired solution in the rules for clarity, as the learning objective does not include there being a difficulty in determining this aspect of the solution in the first place.

Overall, most questions could be answered simply by directing students back to the rules. This emphasises the importance of making sure the rules are clear and represent the examiners intentions well from the beginning.

4.4 Invigilating

Before the exam, the invigilator team inspected the exam room and removed any rule violations found. Examples of actions taken include erasing writing on the whiteboard, and finding and disposing of paper sheets with 59 digits of hidden throughout the room. Students were then let in and told to find a seat, but deliberately invigilators didn't tell them to sit down. We also inspected items brought by students to make sure they did not have π written on them in the clear. To mimic a normal exam, we read out the official invigilator announcement for exams at the University of Birmingham, with minor adjustments to avoid contradicting the rules given to students.

Students ran several cheats that the invigilators could not stop, the most notable including a drone carrying a sheet displaying sixty digits of outside the room's window, and the digits of being repeatedly read out through the room's speaker system.

Students were told in the rules that if they were seen breaking the rules by an invigilator they would fail and have to leave, in actuality we decided to enforce this very leniently. Instead, we agreed to ask an offending student to stop what they were doing or put something that broke the rules elsewhere in the room. This way there was ability for them to discuss their reasoning for believing this was within the rules and if we deemed they had interpreted it incorrectly then they still could try to use other options. This meant that those with backup options could attempt them if their 'Plan A' failed. A selection of creative solutions, both photographed during the exam and afterwards submitted by students are shown throughout the paper in Figs. 2, 3, 6, 7, 8, 9, 10.

4.5 Marking and Assessment Setup

We ran this exercise as a summative assessment, but with only a token amount of credits at 2% of their overall module mark. The exam was strictly pass or fail, as students either correctly wrote all digits or not. In our exam all students who participated ended up passing. The relatively small mark value was chosen to encourage students to engage but to avoid the stress a novel and uncertain exercise may induce in students. If ran as a formative assignment we were concerned on losing out on engagement or more interesting solutions, as it "wouldn't matter". We also didn't want to harshly penalise anyone if they did fail the exercise, as the setup itself invites an inherent level of circumstance that could either help a student pass or cause them to fail. We also told students there would be prizes for:

- The most effective method
- The most imaginative method
- The best failed attempt
- The method the TAs liked best

This incentivised students to attempt more creative and sophisticated solutions in spite of potentially more straightforward strategies. We saw a wide range of solutions which we categorise and discuss in more detail in Sect. 4.7.

4.6 Educational Evaluation

After the exam, we asked the students to fill in a feedback survey, of which 68 from the 86 participating students did. Overall we had 70 responses, two of which were students that didn't actually take the exam. This quiz was optional, but we incentivised responses through making it the means by which students could submit their solution details for judging of prizes. To judge how successfully the exercise conveyed the adversarial thinking mindset, as well as whether the students found the experience enjoyable, we asked the following questions:

1. Were the learning objectives of this exercise clear? Do you understand why we ran it and how it fits in with the module?
2. Did you enjoy this exercise?

The responses for these two questions used a Likert scale, the results of which can be seen in Fig. 4. These questions were not mandatory. We had 65 responses to both of these questions, as there were some students that filled out the later questions but did not choose answers for these ones. 83 objectives of this exercise were "Very clear', and there were no recorded responses falling under "Somewhat unclear" or '"Very unclear". There were 5 responses that did not choose to specify how clear the learning objective was. In terms of the second question, the majority of students enjoyed the exam, with 83 the exercise was "Very enjoyable". There were only 7 responses that did not choose an explicitly positive response, 5 of whom didn't provide a response at all. The one "Very unenjoyable" response was given by a student that was unable to take part in the exercise as they were based in the secondary Dubai campus, which has a different module team and setup[5]. As they did not actually do the exercise, this

[5] The written response implicitly indicated disappointment that they were unable to participate in the exam, but it was also an obviously satirical comment so we refrain from removing it from our results.

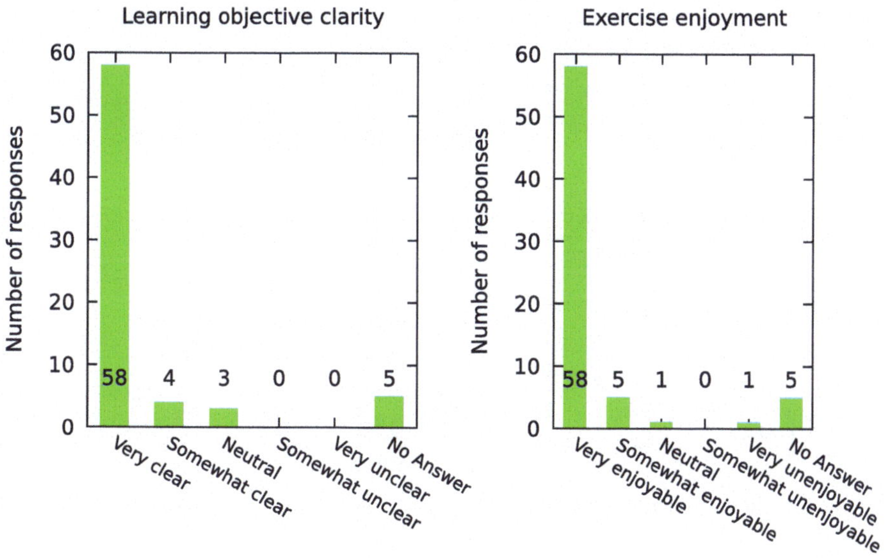

Fig. 4. Student feedback on exercise

specific response does not actually describe an opinion on the exercise. Overall, student opinion shows an obvious broad positive feeling towards the assignment, which matches enthusiasm observed on the day by the invigilators. The module team considered the results to be a resounding success on the part of the exercise.

The next set of questions concerned what solutions the student used in the exercise. These responses were freeform answers, and were not required to be filled out. We asked:

- What method did you use for your final answer?
- Did you try any methods that failed?
- Did you have any backup methods that you didn't end up using? If sop lease tell us what they were.

We used the responses to these questions to create a list of solutions. We found that many solutions followed similar approaches; these have been categorised in Table 1. These are discussed in more depth in the following section. As for the other two questions, 10 students gave meaningful answers to question 2. 7 of these answers described how a solution that they brought into the exercise failed, mostly due to the invigilators' actions. One solution (seen in Figs. 2 and in 7) was inadvertently thwarted by the invigilators turning off the projector screen as part of setup - not realising that someone had modified the display output as an attack. The remaining three answers described ideas that they had considered for their solution, but were then abandoned in favor for another solution. 36 students told us about the backup solutions that they had prepared. One student described as many as 3 backup solutions. Some of the backups described were also just alternative ideas that students had, without having actually prepared it for the day. Broadly, there was a great variety of ideas. We were happy to

Table 1. Categorisation of π exam solutions: survey questions answered by 64 students out of 68 respondents

Class	Final solution		Backup solution(s)	
Loophole	48	75.0%	32	50.0%
Encoding	24	38.5%	11	17.2%
Implicitly allowed device	11	17.2%	13	20.3%
Alternative rendering	6	9.4%	4	6.3%
Memorisation	4	6.3%	4	6.3%
Social	16	25.0%	13	20.3%
Opportunist	9	14.1%	6	9.4%
Teamwork	7	10.9%	7	10.9%
Cheat	8	12.5%	8	12.5%
Covert	5	7.8%	6	9.4%
Hidden in plain sight	3	4.7%	2	3.1%

see the level of engagement students had not only on the day, but also with brainstorming and prior preparation evidenced by the many backups described.

4.7 Categorisation of Student Solutions

As part of our evaluation process, we categorised the solutions described by students. The solutions can be broadly split into three categories. Some solutions fall under multiple classifications, as they meet the criteria for more than one class, though we generally tried to avoid this when possible.

Three quarters of the reported successful solutions methods fell under the loophole category. These were solutions that fit within the constraints of the rules when followed exactly. Within this category were solutions that encoded in an obfuscated way, so that they did not violate the rule regarding having written material (see Figs. 5,6 as examples). Other methods included in this class were devices not explicitly excluded by the rules. Smartphones were not allowed, so instead attacks made use of devices such as E-ink displays, Raspberry Pi Picos, Arduinos, MP3 players, and walkie talkies (which communicate over radio rather than Bluetooth or Wi-Fi). One high effort solution involved bringing in a server (see Fig. 8) to connect to a display in the examination room. Solutions classed under alternative rendering exploit the strict semantic reading of the rule about presentations of that were not 'written' or 'printed'. Examples of this encountered was being: carved into a piece of card, baked into a pie, sewed into a jacked sleeve and painted as a mosaic. A small amount of students also memorised all required digits of π. Some had already had some amount of digits memorised before the assignment was issued, and some memorised it specifically for this assignment.

Some solutions relied on the actions of other students. In this category were people that planned ahead of time to work together for their solution, such as a group that coordinated a drone flight outside with a poster hanging from it. A solution that could have only worked with a group, but also was classed

Fig. 5. A simple encoding using colours on a bracelet divided into sets of five. The first ten colours are the key used in decoding.

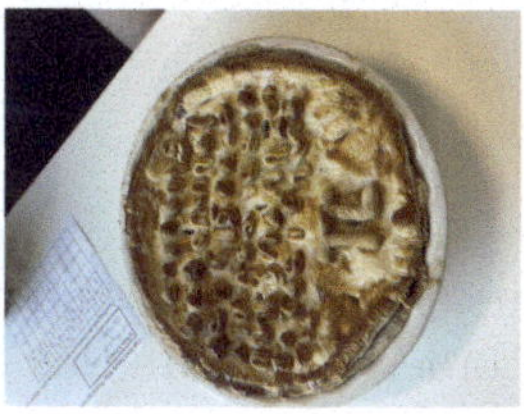

Fig. 6. A more complex encoding baked onto a pie. In this case the key is the number of edges each shape has. For obvious reasons this was a favorite amongst TAs.

under memorisation, were six students that assigned each other 10 digits of π to memorise, who then collaborated together in the exam. There were also the students that made the decision to go in without any solution, believing that the other students' solutions would provide an opportunity for them. This is a high-risk strategy, but in many cases this worked very well for these opportunists.

Some solutions we classify as *cheats*: these would be any solution that violates the rules set, however if students could get away with it they were awarded marks. If these solutions were discovered by invigilators, the student would have failed. These make up the smallest category, but were impressive as attacks because of how well they worked. The success of these solutions were based on how well they could go undetected. A small amount of these were "hidden in plain sight", using explicitly allowed items with slight modifications to hide π completely in the clear. The Figs. 9, 10 show examples of such solutions. The other cheats relied on being hidden and undetectable. One such example is where a student had written in a transparent pen in a way that the text could be hidden by their hand, and also having a second pen as 'a distraction'. Another example utilised an Arduino device (see Fig. 2) - technically allowed as a device, but the student makes a point of mentioning how they tuned it to be 'tactile but completely silent'[6].

When discussing potential solutions as part of the planning process, encodings was not something we considered as an obvious solution, so the end result where the majority of solutions incorporate some manner of encoding was novel.

[6] "... I could have it in my hoodie pocket and run it completely unnoticed, so even though it does technically fit in the rules I was guaranteed to be good. However this discreetness wasnt needed with how obvious everyone else was."

It appears to have been an approachable, simplistic way of enciphering the information just enough so as to not break any rules, but still allowing for a good amount of creative expression and variety in student attacks. Thinking about the learning objective of this specific exercise, we ideally wanted to see more loophole style attacks than cheat style attacks. These loophole style attacks can be related to e.g. traditional protocol attacks, where oversights in design allow for exploits. Such protocol attacks and other similar cybersecurity attacks feature heavily in the module this exercise was run for. That the majority of students used a loophole as their main solution indicates good understanding of the learning objective, marking the exam as a successful teaching method.

4.8 Effect on Prospective Students

In the run-up and the aftermath of running this exercise, the excitement around the module spread outside our module through to other students in the Computer Science course. In order to explore this effect, and whether it made students eager to choose the module, we sought out prospective students in the process of deciding which modules to take in their final year.

We used Discord to gauge the opinion of prospective students - these are students that would be more likely to be talking to those from other years. One of the TAs asked questions in an open forum (an open Discord channel), and received six distinct responses from students considering enrolling in the course. While we had done nothing to inform these students of the existence of this exercise, 3 of the 6 stated that they had already heard of it from speaking to students in the year above. They stated that it had made them more likely to take the course. We found that many students were already aware of this from speaking to students in the year above. There was a generally positive sentiment among these students, who were already excited by the idea of participating in this assignment. We did not receive any negative comments, or comments that indicated that they are dissuaded from taking this module due to this assignment. One student said that they "think it sounds great and was part of the reason [they] picked [the module]", another that "it made [them] more likely to

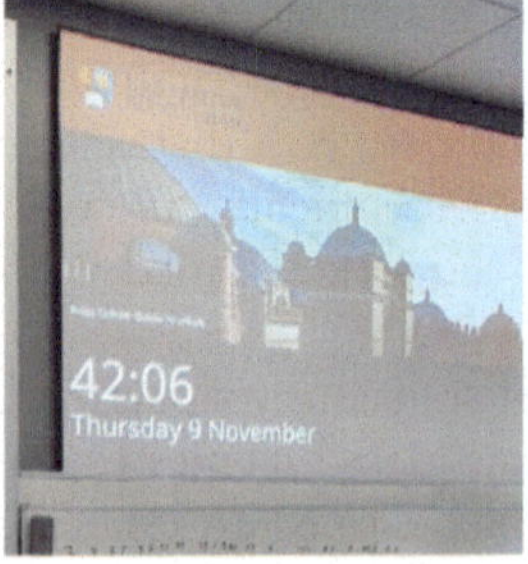

Fig. 7. Student submission: "TLDR: I created a fake login screen that would display the digits of in the time every 5 minutes after 11:20am." Shut down by the invigilators turning the projector off for unrelated reasons.

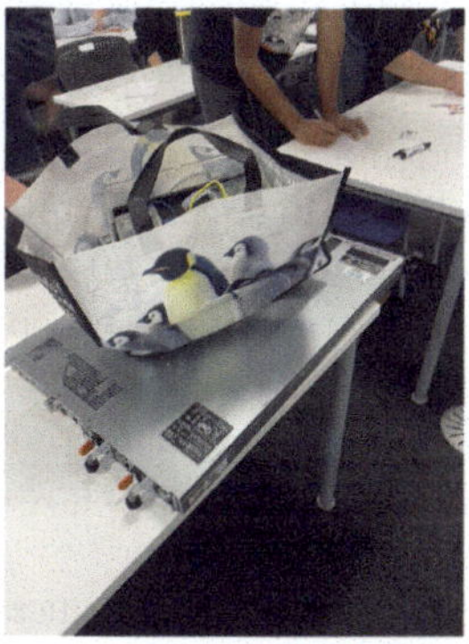

Fig. 8. A server along with the required cables and peripherals to use. Nothing in the rules forbids such a device from being used.

pick [the module]". One student said that "it's unlike any other exam I've heard of", " ...here's this set of constraints, now effectively cheat without breaking them ... it just sounds really fun". These comments then generated further discussion for those interested in this module and what their own solutions would be. These responses are reassuring that students would still engage well with the upcoming assignment, and serve as a promising start to in our efforts to gauge if the exam has a significant effect on module takeup.

These results provide promising indicators that our design has academic merit. and that it would be worthwhile to run the exercise again in some form. The module Security of Real World Systems received high praise this year on the National Student Survey (NSS), being described as engaging, well run, as well as fun and challenging. While we cannot say which aspects of the module students are referring to here, we believe this exercise had a large positive impact on the module. With future iterations we would like to record a more exact measure of the academic benefits this exercise gives students.

5 Discussion

5.1 Exam Question

In running our first iteration of this exercise, we chose to have students recite the first 60 digits of π; this was derived from Conti and Caroland's exercise in which students were required to recite the first 100 digits of π. We chose to shorten the required answer slightly in order to reduce what we saw as an unnecessary complication - we felt students that were able to achieve the required objective of 60 digits of π would have already demonstrated a workable solution and any more would be unnecessary additional work. Additionally, we considered this above the lower bound under which students would consider memorising the required answer rather than finding a solution that better fits the intended learning outcomes.

5.1.1 Memorisation

We feel this worked well, but there is room for improvement. The most obvious concern is the ability for students to memorise the question answer - in our case, this represented a small contingent of students. One student remarked that they had already memorised well over 60 digits of π before knowing the premise of the exam. Additionally, one of the prospective students interested in taking the module also has many digits memorised: "i know like 200 rn i think i could get to 1000 if i tried". We feel that this discourages the use of written prose for answering the question; we believe that prose would be more straightforward to commit to memory.

Of particular note is the use of π for the question. There is something of an understanding that π is a numerical constant the people memorise; in some circles the ability to recite it is a mix of challenge and point of pride. While a very small number of students were already able to recite some or all of the required 60 digits, it is unclear to us if this phenomenon caused an increase in the number of students that more readily reached for memorisation as a solution.

5.1.2 Repeatability

There is some thought to be put toward whether re-running the same question would be beneficial: in our discussions with prospective students they have already started crafting their own solutions. While this is heartening to see it may not be the desired outcome. There is also the possibility of students using previously successful solutions again. It should be first established whether this should be discouraged, or whether it actually is in the spirit of the learning objective (in the same way opportunism was a valid solution for various students). Generally, we want students to come up with novel solutions in order to understand the adversarial thinking mindset, so variety in the question given may prevent these potential issues.

5.1.3 Encoding

Another consideration is the ease of encoding and decoding the answer; in this case, encoding a series of numerals is fairly straightforward. Using a string that is more difficult to encode straightforwardly - consider a string with a wider range of characters in use - would discourage encoding solutions or demand that they be more sophisticated.

5.1.4 Novel Questions

Core to this premise is that students are told the question far in advance; however this too can be modified to adjust the difficulty of the exercise. Consider also a similar exercise where the question is not revealed beforehand. This demands solutions be both online and bidirectional - offline solutions such as encodings are completely eliminated and any solutions must also allow the student to communicate the question as well as receive the answer. This would raise the difficulty substantially. Another question in this vein with the same premise could be to have an established text or source of knowledge (e.g. a novel) that could not be present in the exam venue, and in the exam have the previously unrevealed question ask for a specific subsection of that text (e.g. "Give me the 4th paragraph from Chapter 7"). This question would allow for offline solutions like the question does, but having effective and easily enciphered offline solutions would be significantly more challenging. Conversely, this style of question could encourage attacks requiring teamwork, where the goal is to then communicate to the student in the exam venue the correct answer. It is worth considering how much the question itself affects the solutions that are used to attack it.

5.1.5 Configurability

The main concern of the question asked of students is that it serves to set the difficulty of the exercise and in what dimensions it is difficult. Consider if encodings should be encouraged or discouraged; also consider if multiple questions could be asked to define a moving level of difficulty. For example, by asking one question known in advance and one novel question, it is possible for solutions to be rewarded differently based on their complexity.

5.2 Opportunistic Student Behaviour

While it did not represent a significant proportion of the solutions, about 14% of solutions were 'opportunistic', which is to say that these students instead relied on eavesdropping a solution from another student.

Some of these were cases where the student did not originally intend to do so, but made do when their initial solution failed; others were students that

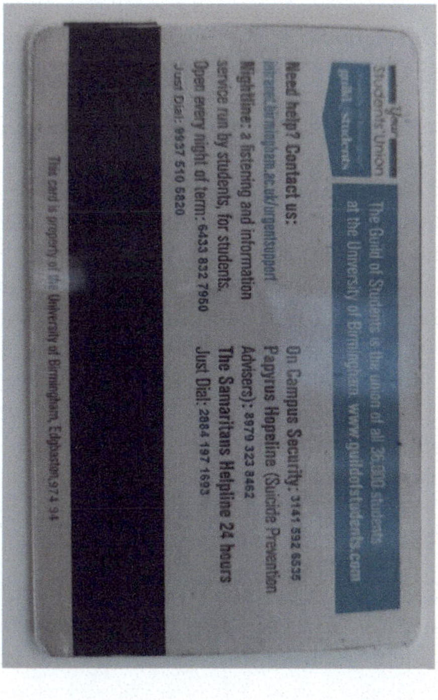

Fig. 9. Here the student has printed out a fake back to their ID card, replacing all the numbers with the digits of π. This breaks the rules but went undetected.

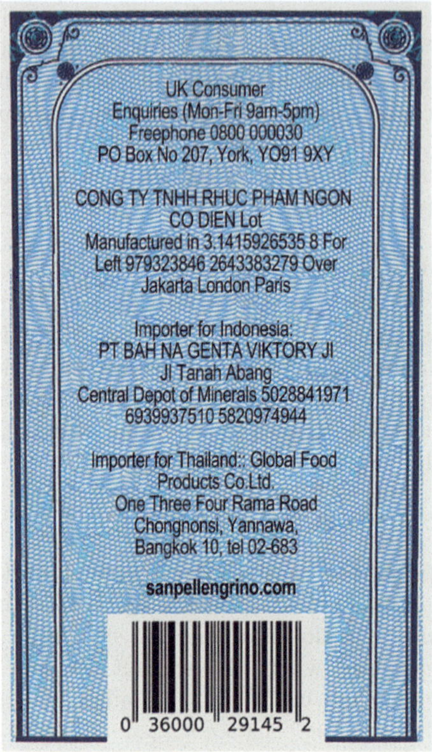

Fig. 10. Another hidden solution, the student has replaced parts of the text on this drink label with the digits of π. Again, this breaks the rules but remained undetected.

entered with no plan except that they would be able to leverage another student's solution.

It is unclear how desirable this behaviour is; ideally all students would have to engage in the act of examining the rules and creating solutions. However, this does appeal to the naturally social aspect of the exercise - which we feel is important - and it is unclear exactly how to discourage this behaviour without also impeding teamwork-based solutions. In running the exercise we found it an acceptable solution; we considered it a high-risk strategy. Important to note is that this is partially modulated by the chosen rules - we chose to allow students to walk around (should they notice that specific wording in the rules), which allowed more opportunistic strategies. Conversely, teamwork allows more sophisticated solutions; we feel that this should be encouraged.

5.3 Arbitrating Rule Clarification

As mentioned, it is essential to maintain an open dialogue with students between releasing the rules and the exercise day. This channel should be accessible to all students - if students contact any member of the exercise team individually, any clarifications beyond being told to re-examine the given rules should be announced to all students.

These clarifications should also be centralised so that they can be referenced by members of the exercise team before responses are given; in our case confirming every "ruling" or clarification between the team was infeasible, so every member of the team had the authority to make these clarifications.

To emphasise the importance of centralising the results of these clarifications is highlighted well by a case we encountered. We were not keeping track of these clarifications strictly enough, and one member of the team - also unable to attend to invigilate on the day - told a student that a specific device, a game console, would not be in breach of the rules. However, on the day, this was determined in breach of the rules as it could communicate over WiFi; this unfortunately meant that the student was advised in one way but the rules were enforced in a conflicting way.

While this was the only case of this we received in feedback, it demonstrates the importance of clearly and consistently arbitrating rule clarifications.

5.4 Student Incentives

In our iteration of this exercise we used two incentives.

The first of these was an academic incentive; we made the exercise contribute a small token amount to the overall module grade for students. We did this to increase engagement as we believed that the students would be more likely to participate in a summative exercise, despite its relatively small contribution to their grade; although this it's actual effectiveness in this role is not clear. The downside to making this assessment summative is the potential for student discontent; the exercise is unlike many they may have undertaken before, and perceived unfair rule enforcement - for example - is more likely to cause upset. We see no reason that this exercise cannot be ran formatively, especially in scenarios where very high student engagement is common and expected.

The second was a set of small prizes, described above. These served the secondary purpose of encouraging students to engage creatively, and for this purpose we feel that they worked well. They also served to incentivise students to describe their methods to us, which is the only means by which they could be entered to win such a prize. Anecdotally, we found that students were excited to share their solutions with us; many of those that employed sophisticated solutions were proud of them and happy to demonstrate them in exchange for the potential of winning a prize.

6 Future Work

While we believe the exercise was a success, we have also identified a number of possible future directions. First, is re-running the exercise with a different question, as discussed in Sect. 5.1, to encourage a wider variety of solutions. Most notably, we would increase the difficulty of encoding the answer. Also, we have considered re-framing part of the objective by introducing a second, unseen question, possibly with a higher mark available. This way students will have to more thoroughly break the system in order to complete both challenges. This second question would also not be achievable via an encoding, encouraging more novel and complex solutions.

In addition to alterations we may make to the exercise itself, we are also interested in further educational evaluation. Notably, more substantial data collection and analysis, including a comparison of solution complexity and overall module attainment. A key purpose of the exam was to get students to think creatively and increase engagement, therefore we are interested in measuring the effect this exercise has on engagement with regular course content in addition to student attainment.

7 Conclusion

In this paper we have described our live security exercise in adversarial thinking and creating problem solving, run at the University of Birmingham, UK. We have outlined the attacker model given to students, as well as the rules developed to encourage the adoption of complex and novel solutions. We have analysed student feedback, finding that most students both clearly understood the learning objectives, and enjoyed the exercise. In addition to analysing student feedback, we have also categorised solutions and outlined our thoughts on each. Finally, we have discussed some of the key lessons learned from running this exercise, with the hope that they will help advise other educators on how they might best deploy this exercise.

Appendix

Exam Rules

The exam rules we gave to students are as follows:

There will be one question on the exam, which is: write out the first 60 digits of pi.

The exam will be location on date. It will start at start time and end at end time i.e., a 40 min exam. No one will be allowed to enter after the first 5 mins, or leave before the last 20 mins. TAs will escort students to the rest rooms if needed (but only if there are TAs available for this).

You may not bring in anything that has been written or printed on (not including e.g., brand names on drinks, clothing, etc.). We will check the room before the exam and will remove any writing that we can identify to encode the first 60 digits of pi.

No laptops, tablets, smartphones or smart watches. No calculators or any device that readily displays stored digits of pi. You must not use Wifi or Bluetooth wireless connections.

Once the exam has started, students in the exam room must not talk to anyone else in the exam room, except the invigilators.

Students may not pass anything to each other during the exam.

Students must bring their ID card and place it face up on their desk.

Students may choose their own seats. Once seated, they must stay in their seat until an invigilator gives them permission to leave. If you need to leave, raise your hand and an invigilator will come to you.
Any student seen breaking the rules by an invigilator will have to leave the exam hall and will fail on the test.
At the end of the exam, we will collect the exam papers from the desk. There is a box to write your student ID number and name on the paper. The students whose ID and name is in that box will get full marks if the paper has pi to 60 digits on it.
You must follow all instructions from an invigilator, in particular, you must leave the exam room when told to do so by an invigilator.
Any attempt to deliberately disrupt the exam for other students is strictly forbidden.
Note: you are not actually expected to memorise the digits of pi. The key learning objective of this module has been to think like an attacker, i.e., the ability to look at a system, learn its rules and find a way to exploit the system.
The rules of this "exam" will be enforced strictly and to the letter. There are no other rules. All the normal expectations of good, honourable behaviour that are required for the other exercises on this module, and other modules, are not required here. The normal de- scription of academic misconduct and examination irregularities in the student handbook do not apply. If the rules outlined above are deemed to be broken, you will only fail this test, it will not be treated as a genuine case of an examination irregularity.
Marks will be assigned based on the exam papers. After the exam you will be invited to fill in a brief Canvas quiz to tell us how you approached this exercise. There will be prizes for: 1) The most effective method, 2) The most imaginative method, 3) the best failed attempt and 4) the method the TAs liked best.

References

1. Carlisle, M., Chiaramonte, M., Caswell, D.: Using CTFs for an undergraduate cyber education. In: 2015 USENIX Summit on Gaming, Games, and Gamification in Security Education (3GSE 15). Washington, D.C.: USENIX Association (2015). https://www.usenix.org/conference/3gse15/summit-program/presentation/carlisle
2. Chothia, T., et al.: Choose your PWN adventure: adding competition and storytelling to an introductory cybersecurity course. In: Transactions on Edutainment XV, pp. 141–172 (2019)
3. Conti, G., Caroland, J.: Embracing the Kobayashi Maru: why you should teach your students to cheat. Secur. Priv. IEEE **9**, 48–51 (2011). https://doi.org/10.1109/MSP.2011.80
4. Davis, A., et al.: The fun and future of CTF. In: 2014 USENIX Summit on Gaming, Games, and Gamification in Security Education (3GSE 14) (2014)
5. Ensafi, R., Jacobi, M., Crandall, J.R.: Gamification in security education (3GSE 14). San Diego, CA: USENIX association (2014). https://www.usenix.org/conference/3gse14/summit-program/presentation/ensafi
6. Students who don't understand information flow should be eaten: an experience paper. In: 5th Workshop on Cyber Security Experimentation and Test (CSET 12). Bellevue, WA: USENIX Association (2012). https://www.usenix.org/conference/cset12/workshop-program/presentation/Ensafi
7. Vigna, G.:Teaching network security through live exercises. In: Security Education and Critical Infrastructures. Ed. by Cynthia Irvine and Helen Armstrong. New York, NY: Springer US, pp. 3–18 (2003). ISBN: 978-0- 387-35694-5

EXCITE: EXploring the enhancement of Cyber security training through Immersive TEchnology

Sandra Scott-Hayward(✉), Michelle Butler, and Carole Parsons

Queen's University Belfast, Belfast, UK
s.scott-hayward@qub.ac.uk, michelle.butler@qub.ac.uk, c.parsons@qub.ac.uk

Abstract. The goal of the EXCITE (EXploring the enhancement of Cyber security training through Immersive TEchnology) project was to explore the potential to increase the efficacy of cyber security training for third-level students through the use of immersive experiences supported by virtual reality design. In the first phase of the EXCITE project, we explored the cyber security experience of a diverse, multi-disciplinary group of third-level students, their views on the suitability and effectiveness of cyber security training materials, and their perspectives on the potential use of immersive technologies in cyber security training. Based on our findings in Phase 1, Phase 2 focused on the design and development of a suite of cyber security immersive experiences presented in virtual three-dimensional student-oriented scenarios e.g., student residence, university computer lab, café. The experiences cover the topics of *Password Guidance, Phishing Awareness, Social Engineering, Your Digital Footprint, Social Media Use, Secure Remote Working, Protecting Your Data and Devices,* and *Incident Reporting.* In this chapter, the design of each phase of EXCITE is presented, along with its findings, feedback from students, and recommendations based on the experience of rolling out the cyber security awareness experiences in a university.

1 Introduction

Universities are key contributors to the economy, skills development, and innovation in the UK. In making this contribution, they handle personal and research data, intellectual property, and other assets, each of which has significant value to others. Further, those individuals that come to university learn the foundational skills, across all disciplines, upon which they build their careers; this represents a significant opportunity to bring cyber security knowledge to a broad spectrum of industries and roles to strengthen digital security for years to come. In the EXCITE project, the potential to increase the efficacy of cyber security training for students through the use of immersive experiences supported by augmented and virtual reality (AR/VR) design were explored.

There were two main motivations driving EXCITE. Firstly, a gap in the provision of cyber security awareness training that is broadly accessible to all

P. Legg et al. (Eds.): CSE 2024, LNNS 1213, pp. 96–115, 2024.
https://doi.org/10.1007/978-3-031-77524-6_6

third-level students of any discipline was identified. Secondly, traditional Massive Open Online Course (MOOC) style training materials can be very one-dimensional with, for example, a student working through a slide presentation and then taking a short assessment/quiz to demonstrate knowledge of the presented topic. This often results in a "box ticking" exercise, which limits the student's engagement and the efficacy of the training. Additionally, in a real-world setting, there are often distractions that can influence actions or reactions to a cyber security situation, but this real-world experience cannot be replicated in traditional MOOC style training materials.

EXCITE aims to address these challenges by helping third-level students, from across disciplines, to better understand cyber security, develop their knowledge and skills in this area, and enhance their understanding of the potential consequences of poor cyber security both personally and professionally. In the first phase of the EXCITE project, the cyber security experiences of a diverse, multi-disciplinary group of third-level students, their views on the suitability and effectiveness of cyber security training materials, and their perspectives on the potential use of immersive technologies in cyber security training were explored. Based on the findings from Phase 1, Phase 2 focused on the design and development of a suite of cyber security immersive experiences presented in virtual three-dimensional student-oriented scenarios (e.g., student residence, university computer lab, café). Two versions of these experiences were developed; a desktop application available for students to download and run on their own device to support broad accessibility of the training, and a VR headset application to provide a fully immersive environment for focused, small group training.

In this chapter, we present the design and development of the EXCITE cyber security immersive experiences and our findings based on testing with student focus groups during the development phase, open sessions with our university students and staff to try out the final VR experiences, and the 3D application roll-out across our university. We offer recommendations to increase accessibility and adoption of EXCITE as higher education cyber security awareness training.

2 EXCITE Phase 1

EXCITE Phase 1 was an interdisciplinary cyber security project led and supported by members of the Queen's University of Belfast (QUB) Academic Centre of Excellence in Cyber Security Education (ACE-CSE) team (Dr. Sandra Scott-Hayward (Faculty of Engineering and Physical Sciences (EPS)), Prof. Michelle Butler (Faculty of Arts, Humanities and Social Sciences (AHSS)), and Dr. Carole Parsons (Faculty of Medicine, Health and Life Sciences (MHLS)) working with students from across the university and an external advisor company, Digital Catapult NI Immersive Labs, to review and analyse traditional cyber security training materials[1] and explore the design of an immersive training experience.

2.1 Study Design

The study design consisted of questionnaires and focus groups, to capture the students' knowledge of cyber security and their thoughts regarding the cyber security training materials and how immersive technology could be used to enhance this training. Ethical approval to conduct the study was obtained from the QUB EPS Faculty Research Ethics Committee. The study was centred on student participation with their activities separated into four different stages.

2.2 Student Recruitment Process

Twenty-five students were recruited to participate in Phase 1 of EXCITE, drawn from across the three University Faculties and involving postgraduate and undergraduate students. Eleven students were recruited from AHSS, six from EPS, and six from MHLS. Fifteen of the twenty-five students were postgraduate taught students, six were undergraduate students and two were postgraduate research students. Students were deliberately recruited from all three Faculties at undergraduate and postgraduate levels to ensure a diversity of perspectives.

Potential participants were voluntarily recruited using a convenience sampling approach, using the following inclusion and exclusion criteria:
Inclusion criteria:

- aged 18 years or over
- able to read and understand English
- capacity to give voluntary, informed consent to participate in the research
- student at QUB

Exclusion criteria:

- aged under 18 years
- unable to read and understand English
- lacking capacity to give voluntary, informed consent to participate in research
- not a student at QUB

2.3 Study Stages

The study involved four different stages, all of which involved student participation. A description of each stage is provided below.

Stage 1: Students completed a questionnaire about their prior training/knowledge of cyber security at school, university, on work experience and/or

[1] We define traditional cyber security training materials as those that are one-or two-dimensional. An example of one-dimensional training is a student working through a slide presentation and taking a short assessment/quiz to demonstrate knowledge of the presented topic. An example of two-dimensional training is a student watching a video role-play and taking a short assessment/quiz to demonstrate knowledge of the presented scenario.

in current or previous employment. For example, students were asked about what topics any training had covered, their use of passwords, their recognition of phishing from example emails, their awareness of QUB IT and cyber security policies, and their awareness of incident reporting mechanisms.

Stage 2: Students individually reviewed a set of allocated traditional cyber security training materials[2] and completed a questionnaire relating to the provided materials. In addition to questions reviewing the content, such as whether it was pitched appropriately, felt effective and relevant, and if they learned anything from it, students were asked how long it took to work through the training material and what device they used to interact with the materials.

Stage 3: In this stage, students attended a focus group to discuss their feedback on the provided cyber security training materials and completed exercises to measure their recollection of these training materials. Each focus group comprised five or six students and was facilitated by an ACE-CSE team member. The students' thoughts about the training were discussed, as well as how the training materials related to cyber security issues as experienced in the real world. In particular, students were asked which training materials stood out for them and why, which materials were best designed for university students and why, how the training materials could be made more effective, and if their cyber security behaviour had changed since reviewing the materials. The efficacy of the traditional training materials was also assessed by asking students to complete the following two cyber security exercises: the National Cyber Security Centre training quiz[3] and the OpenDNS Phishing quiz[4].

Stage 4: Finally, the students attended a second focus group to gain an understanding of immersive technology and training capabilities, and discuss their feedback on the use of immersive technology for cyber security training. In this session, the students were introduced to immersive technology (XR), had a "hands-on" VR demonstration, were shown cyber security training use-cases featuring XR as a tool and then provided feedback on their views regarding the potential use of XR to provide more effective and tailored cyber security training to university students. There were five students in each session led by Digital Catapult NI.

A. Introduction to XR. The session started with a short presentation on the context to XR, its associated terminology and applications, the aim of which

[2] NCSC (Top Tips for staff), CPNI (Social Engineering, Don't Take the Bait! and It's OK to Say), QUB Staff (Email Essentials, Essential Phishing Awareness, and Secure Remote Working), SANS Secure the Family (Top Threats to the Home, Securing Devices, and Backing Up Your Digital Information), and Open University (Securing my digital information, Improving password security, How malware gets into your computer, and What's the worst that could happen?).

[3] https://www.ncsc.gov.uk/training/top-tips-for-staff-scorm-v2/scormcontent/index.html/.

[4] https://www.opendns.com/phishing-quiz/.

was to increase the students' understanding of and familiarity with the technology to help them express their ideas or opinions relating to cyber security awareness training. The main differences between Augmented Reality (AR), Virtual Reality, and Mixed Reality were discussed, as well as recent advances in Head Mounted Displays. The presentation also included explanations around eye tracking, pass-through camera technology and how hand gestures can be utilised in XR as a navigation and user interface (UI) system.

B. Hands-On VR Demo. A VR experience (First Contact designed by Meta) was selected to ensure that the students would get a wide variety of immersive technology interaction in an engaging virtual environment and learn how hand movements and user agency is a key function. It was deemed important that the students learned the interactive and intuitive nature of using VR equipment - most for the very first time - in a simple but engaging experience that was not specific to a cyber security training example. The benefit of this was to enable the students to experience the sensation of VR and to understand its nuances before considering how it might be applied in cyber security training. Oculus Quest and Quest 2 headsets were used by the students.

C. Cyber security training use-cases featuring XR. Next, the students were presented with a series of video examples from companies using XR to train their staff, which served as a use-case of how XR could be used to provide cyber security awareness training. These included:

- A company using AR and gamification to make their online cyber security training more engaging. For example, representing phishing as a cartoon character who would 'jump out of' an AR mobile device.
- An immersive storytelling approach (e.g. 'A Date with a Hacker'), which puts the user in a VR headset and introduces the various personas that the would-be hacker might deploy to trick someone into revealing sensitive and secure information.
- A VR scenario where the user uses 'gaze' cues to look around their home office to spot potential cyber security threats prior to joining a work-related call.

D. Student Feedback Session. Finally, the students were asked for their feedback regarding the potential use of XR to provide cyber security awareness training to university students. More specifically, they were asked their opinions on whether XR would enhance the learning experience, if VR functions or tools would be useful in delivering cyber security training, if there were any XR techniques that would be beneficial to incorporate into cyber security training, what barriers XR might present in cyber security training, and if there were any particular cyber security situations in which they thought the use of XR would be particularly useful.

3 Phase 1 Results

3.1 Stage 1

The initial survey provided a baseline for the cyber security knowledge/experience of the students before completing the training materials. Despite the diversity of the cohort, the majority reported some previous training and a reasonable knowledge of cyber security. This may be influenced by their voluntary participation in the study indicating their interest in cyber security. However, despite this, some poor cyber security practices were still evident in their reported behaviour. This suggests that the training they had received was of limited effectiveness. Similarly, there was a lack of awareness of what cyber security information was available, how to access it, and what to do in the case of experiencing a cyber security threat/incident.

3.2 Stages 2 and 3

The review and discussion of existing cyber security training materials provided insight into the format and content of such training that might be best suited to third-level students. Building on the findings from stage 1, it was evident that any training provided would need to consider the potential diversity in the students' knowledge of cyber security and their varying digital skills and reflect students' different learning styles. For example, a set of introductory materials could be supported with more detailed content for those students interested in finding out more. Furthermore, while some responses indicated that videos can be an effective training tool with the potential to draw the viewer into a scenario, other students preferred to read the material for faster assimilation.

There was also a clear demand for practical elements to be incorporated into the training. The materials that included quizzes or exercises to test understanding were rated more effective and impactful. Of note was the value attached by the students to the provision of feedback on such practical exercises, where the feedback indicated the correct response and provided some explanatory information. Moreover, the student responses pointed towards the potential benefit in impact and effectiveness that might be achieved with an immersive style of training. For example, *"The questions just appear when you are looking at scenarios. This helps reader to have more real feeling and sense of involving."* and *"Creating a situation and asking the reader how to react in such a situation is an interesting and effective method"*.

Similarly, there was a demand for clear practical guidance/recommendations as to what to do, what not to do, and who to report concerns to. For instance, what password manager/Virtual Private Network (VPN) to use rather than an abstract guide to creating a good password or the importance of secure data transfer. Some students suggested that this might remove the fear factor associated with learning about the importance of cyber security without knowing how to practise good cyber security hygiene.

From a content perspective, the students identified the importance of the material being relevant and relatable to their lives. Scenarios that closely matched their immediate circumstances, such as use of their university email account, passwords for university services, and securing devices were considered particularly relevant. Awareness of social engineering, phishing, good password hygiene, digital footprint monitoring, and security on shared networks, such as public Wi-Fi networks, were also highlighted as relevant.

Additionally, the importance of the materials being engaging and grabbing the students' attention was stressed. Some students suggested the use of simulation to convey the information; for example, setting a task for the student to configure the appropriate security settings on a Wi-Fi router. Again, this pointed towards the potential to increase the value of the cyber security training through immersive techniques.

3.3 Stage 4

Most students found the VR experience enjoyable and saw it as having potential value in creating learning opportunities for cyber security training. Students acknowledged the benefits of reading, listening, watching, and completing quizzes that were evident in existing cyber security training materials but felt that these materials would benefit from the use of XR. Many suggested that a mixed/hybrid approach with XR playing a role in future cyber security learning/training experiences would be beneficial.

When using VR, most students reported becoming quickly familiar with their surroundings, and the controls and functions needed to perform the tasks, as well as with the potential impact of the gameplay on learning and behaviours. It was reported that the act of physically 'doing' or completing tasks in an immersive environment was memorable and impactful. Gamification and task-orientated learning were well received. The potential to use the VR experience to 'play through' scenarios as a simulation to replicate what is required post cyber security attack or in real-life situation was stressed.

Agency and control, self-discovery and self-determination afforded in VR, along with an environment which encourages - and almost demands - focus, while giving the user a memorable and innovative experience emerged as key factors in making VR a useful delivery tool for cyber security training.

4 Phase 1 Findings/Outcomes

The traditional cyber security training materials provided to the students for review were considered, for the most part, to be suitable for third-level students and effective in their messaging. Their impact and influence on behaviour change were dependent on the prior cyber security knowledge of the student and the practical nature of the information/training.

However, the results of EXCITE Phase 1 suggest that a hybrid training approach could be more effective, with more traditional online training methods working in tandem with and complementing a VR cyber security training experience.

To accommodate diverse levels of cyber security knowledge across the student population, the training materials should also span a range of levels. For example, a set of introductory materials could be supported with more detailed content for those students interested in learning more.

4.1 Format/Presentation/Structure

To be most effective, cyber security training materials for third-level students should include:

- **Case studies or characters that bring the scenario to life.**
- **The capability for the student to self-check their knowledge/understanding.** For example, the inclusion of quizzes with explanations provided for incorrect answers to enhance learning.
- **The opportunity to apply the knowledge gained.** For example, by using two- and three-dimensional technology to recreate conditions which mirror real world situations that the student is likely to encounter in their daily life. In particular, XR could be used to provide an opportunity for the student to apply knowledge in distracting and/or stressful conditions.

4.2 Content

Cyber security training materials targeting third-level students should include:

- **Content which is directly relevant and relatable to student life.** For example, the importance of backing up student work. Only eight participants had previously received training on secure remote working. During the Covid-19 pandemic, student engagement with the University became reliant on remote access, and this approach to delivery has continued post-pandemic. As such, the security aspects of remote study/work were, and continue to be, of high importance. In Stage 2 of the study, the students were introduced to training on secure remote working and security at home. While useful, their feedback suggested that the content should be made more relevant by presenting security best practices with respect to day-to-day activities linked to their studies and student life (e.g., living in shared student accommodation)
- **Coverage of the "Digital Footprint".** This stood out amongst the presented materials as particularly impactful, with students identifying the value of being made aware of their digital footprint and its relevance to them due to their reliance on digital services and as heavy users of social media. The impact was highlighted through the behaviour changes described by the students as a result of the training.
- **Coverage of best cyber security practice with social media.** *"We explore a lot of websites and social media sites as students, so it's crucial to be aware of how our personal information might be leaked by simply clicking one button."*

- **Coverage of cyber security at home/in student accommodation.** Some students identified the material they reviewed as beneficial for raising awareness of cyber security with their peers, colleagues and family. During the study, students reported their interest in supporting and sharing learning with family, friends, classmates etc.: *"But the different plan is with this material I can introduce these concepts more easily to my colleagues and my family members"*. This is important as training students provides opportunities to bring cyber security knowledge to the wider community, through the people with whom they interact, as well as the industries and roles that they join post-graduation. The students also demonstrated their awareness of the impact of their cyber security practices on the university. *"Students will often get targeted with phishing ads for promotional offers if you just type in your student details, this can often lead to serious data breaches that affect not only the individual student but the university as a whole."*
- **Specific information for international students** who may be especially vulnerable to cyber security attack due to their lack of familiarity with existing practices in the UK and/or use of multiple devices across different networks and regions.
- **Practical guidance on how to respond to a cyber security incident** or how to apply cyber security best practices. For instance, how to minimise exposure to cyber security attacks, what tools to use (e.g., password manager), what technical actions to take, and what to do if subjected to an attack. Some students highlighted the effectiveness of specific messaging. For example, the message that they should report cyber security concerns or incidents, without fear of negative consequences, was highlighted as addressing a gap in knowledge as they often reported not knowing what to do when they received a phishing email.
- **Training and guidance about use of tools** such as Microsoft Teams, Zoom, TeamViewer etc.

From these findings, it was concluded that it would be beneficial to create a suite of cyber security immersive experiences presented in virtual three-dimensional student-oriented scenarios (e.g., student residence, university computer lab, café) which would include specific targeted content for this demographic. Several specific benefits of the immersive technology approach were identified by the students. The ability to practise, perform tasks, and receive feedback through actions and reactions in the immersive environment were identified as being memorable and impactful. Furthermore, playing through scenarios with a VR experience can better prepare the individual to respond to a real-life cyber security incident. The virtual nature of the training can also remove the fear factor around cyber security and attack/incident consequences, enabling open engagement with the material. Moreover, the immersive environment by its nature forces the individual to focus on the cyber security training by removing real-world distractions.

5 EXCITE Phase 2

This phase of the project built on the findings of EXCITE Phase 1 and was undertaken in partnership with Zubr Virtual Reality[5], an organisation experienced in immersive digital content creation for training/learning. It involved creating immersive experiences to provide cyber security training to third-level students and consulting with QUB students to assess the appropriateness of the content, design, and usability of these immersive experiences. Key to EXCITE Phase 2 was ensuring that the design and development of the immersive cyber security training experiences met the requirements identified from the findings of EXCITE Phase 1 presented in Sect. 4. This mapping is detailed in Table 1.

5.1 Study Design

Like Phase 1, the study design for Phase 2 consisted of questionnaires and focus groups to capture the students' feedback on the design, content and usability of the immersive cyber security training experiences being created. Phase 2 consisted of three stages, with students' participation being central to each stage. Ethical approval to conduct the study was again obtained from the QUB EPS Faculty Research Ethics Committee and all student participation was voluntary. As before, students were recruited from across all three University Faculties (AHSS, EPS and MHLS), using the same inclusion and exclusion criteria as in Phase 1, to ensure a diversity of views and experiences. As the number of students participating in the different stages of Phase 2 varied, an account of the number of students participating in each stage of Phase 2 is provided below.

Stage 1: An initial ideation/design workshop was hosted at QUB with participation from QUB and Zubr EXCITE project team members and members of the QUB ACE-CSE core and extended teams. The ideas for immersive experiences for cyber security awareness training topics in *Password Guidance, Phishing Awareness, Social Engineering, Your Digital Footprint, Social Media Use, Secure Remote Working, Protecting Your Devices, Cybercrime Awareness,* and *Incident Reporting* were discussed. Following the workshop, the EXCITE project team discussed potential versions of the immersive experiences to produce a set of storyboards for presentation to the students.

The students were asked to individually review the storyboards, complete a questionnaire collating their feedback on the content, design, and appropriateness of the storyboards, and to attend an online focus group to discuss their responses and suggest amendments that could be made to improve the storyboards. Each focus group comprised five or six students and was facilitated by the EXCITE project team members and Zubr. Thirteen students completed the questionnaire (eight students from EPS, four from AHSS and one from MHLS). Eight were postgraduate taught students, four were undergraduate students and one was a postgraduate research student. Eleven students completed the focus group (seven from EPS, three from AHSS and one from MHLS). These included

[5] https://zubr.co/.

Table 1. Mapping of Phase 2 design requirements from Phase 1 findings

Requirement Description (Phase 1 Finding)	Phase 2 design element
Case studies or characters that bring the scenario to life.	✓ Use of QUB locations - library, café, student room. ✓ Use of QUB branding within these scenarios e.g., posters, merchandise etc. ✓ Use of student characters in relevant scenarios e.g., *Social Engineering, Social Media Use* etc.
Capability for the student to self-check knowledge/understanding.	✓ Inclusion of topic-related question pre- and post-experience. ✓ Gamified experiences to practice and then test knowledge/understanding e.g., *Password Guidance, Phishing Awareness.* ✓ Inclusion of feedback at the end of each gamified experience.
Opportunity for the student to apply knowledge.	✓ Gamified experiences to practice and then test knowledge/understanding e.g., *Password Guidance, Phishing Awareness.* ✓ Use of QUB locations - library, café, student room. ✓ Use of student characters in relevant scenarios e.g., *Social Engineering, Social Media Use* etc. ✓ Timed experiences to create an element of pressure.
Content which is directly relevant and relatable to student life.	✓ Use of student characters in relevant scenarios e.g., *Social Engineering, Social Media Use* etc. ✓ Inclusion of devices/services used by students throughout the experiences e.g., Canvas/Microsoft Teams/WhatsApp etc.
Coverage of the "Digital Footprint".	✓ Specific experience (*Your Digital Footprint*) focused on this topic.
Coverage of best cyber security practice with social media.	✓ Specific experience (*Social Media Use*) focused on this topic.
Coverage of cyber security at home or in student accommodation.	✓ Inclusion of a student room scenario. ✓ Experiences linked to cyber security in student accommodation e.g., *Social Engineering, Protecting your Data and Devices.*
Specific information for international students	✓ Specific experience (*Social Engineering*) focused on this topic. ✓ Services used by international students included e.g., WeChat.
Practical guidance as to how to respond to a cyber security incident	✓ Specific experience (*Incident Reporting*) focused on this topic. ✓ Feedback, recommended tools and best practices included in all experiences.
Training and guidance about use of tools	✓ Information shared on completion of training to cover links to guidance on use of common QUB tools.

a mixture of undergraduate, postgraduate taught and postgraduate research students, including seven postgraduate taught students, three undergraduate students and one postgraduate research student.

Stage 2: The student feedback from Stage 1 was reviewed and discussed by the project team and Zubr and a set of changes and improvements to the immersive experiences was agreed. Three of the proposed immersive experiences were then developed by Zubr - *Password Guidance, Phishing Awareness* and *Secure Remote Working*. QUB provided all the cyber security content to be presented in the experiences, including QUB branding and images. The students were provided with video recordings of each of the three experiences and a short instruction sheet, and asked to complete a questionnaire for each experience to provide their feedback. Sixteen students completed the questionnaire, consisting of six from MHLS, five from EPS and five from AHSS. Six were postgraduate taught students, six were undergraduate students and four were postgraduate research students. These students were also asked to attend a focus group to discuss their responses to the questionnaire and suggest revisions that could be made to enhance the immersive experiences. Each focus group comprised five or six students and was facilitated by EXCITE team members and Zubr. Thirteen students took part in the focus group with six from MHLS, five from AHSS and four from EPS. There were six postgraduate taught students in the focus group, four postgraduate research students and three undergraduate students.

Stage 3: The student feedback from Stage 2 was reviewed and discussed by the project team and Zubr, with a set of changes and improvements to the immersive experiences agreed based on this feedback. The full set of VR immersive experiences was then developed by Zubr. QUB again provided all the cyber security content to be presented in the experiences, including QUB branding, images and audio/video clips.

Finally, students were recruited to participate in a focus group to review the full set of VR immersive experiences developed and provide feedback on these experiences. Each focus group comprised four or five students and was facilitated by EXCITE team members and Zubr. During these focus groups, each student was paired with an EXCITE team member to provide support with the VR headset and instructions for progressing through the immersive experiences. The students were asked for feedback after completing the experiences. Nine students took part in this focus group, with four drawn from MHLS, three from EPS and two from AHSS. These students included undergraduate and postgraduate students (four postgraduate taught students, three undergraduate students and two postgraduate research students).

6 Phase 2 Results

6.1 Stage 1

The students found the storyboards to be mostly appropriate and relevant to cyber security training for third-level students and were excited to see the developed immersive experiences. However, they also made several suggestions for amending or enhancing the storyboards. The main points from their feedback from the questionnaires and focus groups included:

- Use QUB-specific images for the student room/library/café settings rather than stock images to support student familiarity with the spaces.
- Increase time allowed for the phishing awareness games.
- Add digital trophy/badge to the completion of each scenario.
- Place the cyber security statistics near the relevant portal rather than in a separate lobby/quad area.
- Add the "expected time to spend/for completion" at the beginning of each scenario.
- Provide a help option. This should be a button to bring back the instructions screen.
- Include visual cues/audio cues to accommodate accessibility requirements.
- In the *Your Digital Footprint* experience, add animations to demonstrate that the data reaches outside the room and is publicly visible/available.
- In the *Social Media Use* experience, add an example and explanation between social and professional posts to help get the students started. Also, use the example list of websites provided by the students in the focus groups and add in some "never disclose" items to the board (i.e., if the student tries to move these items into either the social/professional circle, there will be a buzzer and pop-up to highlight that they should never disclose this on social media).
- Include the ability to email the additional resources mentioned during the games to the user at the end of the experience.
- In the *Phishing Awareness* experience, add the "importance of reporting" feedback to the phishing scenario so students understand that reporting is useful. Adjust the feedback/introduction to the scenario to account for this.
- For all scenarios, ensure that there is an explanation of what should/could have been done. This could be included as audio at the end.
- In the *Social Engineering* experience, provide an alternative path through this experience.
- Add Internet of Things devices to the student room.

6.2 Stage 2

Student feedback after reviewing the three immersive experiences developed following the results of Stage 1 was largely positive, with students commenting that they found the experiences to be engaging and impactful. Nonetheless, the students provided some suggestions for improvement, including more general design and development enhancements, as well as those relating specifically to each experience. A list of these suggestions is provided below:

General Suggestions:

- Computer generated voice was not engaging. A human voice would be preferred over an Artificial Intelligence (AI) voice. It would be acceptable to have some use of the AI voice throughout the experiences, but the preference is for a human voice to be used predominantly.

- A "gamified" user interface option preferred to a specific QUB branding-aligned option i.e. QUB colour palette/font etc.
- Increase font size for accessibility throughout.
- Add help/tips option that the user can bring up at any time during the experience- can be a listen/read again option rather than specific tips.
- Maintain current design of timer but highlight it with game instructions when linked to the scoring of the experience.
- Provide a mini example with explanation or simple version of the game to begin each experience (where appropriate). In places, this will just be an adjustment to provide a visual illustration to accompany the audio introduction.
- Suggestion that at the end of the game, there could be a link to enter an email address to receive feedback. This suggestion was made for the Desktop version of the experiences rather than VR. For VR, it was suggested that this can be provided directly to students when they have completed the VR training.

Suggestions for the Password Guidance Experience:

- Some cards at the end of the password game and tip boxes during the game were too text heavy - summarise where possible.
- Add option to more closely inspect the password guidance tips.
- Add option to read the introduction.

Suggestions for the Phishing Awareness Experience:

- Add visual aid during the initial information rather than just audio track e.g., keywords that are spoken to appear on a whiteboard.
- Increase the time for the user to review each phishing email; part 1 - increase from 7 to 10 s and part 2 - increase from 10 to 15 s.
- Find another way to present the website address - perhaps providing an information leaflet or email once the experience is completed.
- Give examples first and then practice.
- Include a written version of the feedback at the end for students to read - concise format.

Suggestions for the Secure Remote Working Experience:

- Some information boxes were too text heavy.
- Introduction audio was too short, provide clearer scenario explanation/instructions.
- Provide clearer guidance for which area of the room the person needs to visit to get instructions.
- Add feedback during the walk-around. For example, read out from the textbox as you discover each device rather than saving it all until the end. Speaker to read the prompts at each device.

- Add feedback on physical and personal security at the end e.g., if student card/notebook is missed.
- Include suggestions/links on how to download or use a VPN.
- Add hints after a certain time.
- Sum up findings at the end.
- Identify how many devices/things to spot by adding a progress visual.
- Add the missing explanation on the use of OneDrive.

6.3 Stage 3

Following the review of the full set of immersive experiences, student feedback was very positive. Overall, the students reported that the experiences were fun, enjoyable and "exceeded their expectations". No major changes were proposed to the design at this stage, but some suggestions were made to adjust the audio/video, address some bugs that had been identified, and enhance the user interface and game instructions. These suggestions were discussed and enacted where appropriate by the EXCITE team members and Zubr.

7 Summary of EXCITE VR Immersive Cyber Security Experiences

A brief description of the final full set of immersive cyber security experiences developed in the Phase 2 study for third-level students is provided here along with illustrations provided in Figs. 1, 2, and 3.

Password Guidance. This scenario takes place in the QUB library. An overview of strong and weak passwords and best practice in password creation and usage is presented. The student completes three games to identify strong versus weak passwords. The final game is timed and scored. After each game, the correct ranking of the passwords is shown, and feedback is provided to reinforce the password guidance. The concluding message is to take the password pledge[6].

Phishing Awareness. This scenario takes place in the QUB library. The introduction outlines the tell-tale signs of phishing emails. The student then completes two games to distinguish between phishing and benign emails. In the first game, the student is timed in identifying the emails. In the second, the student highlights the indicators of phishing in the emails. The concluding message provides a reminder about how to protect against phishing attempts, directs the student to the QUB reporting email address and reiterates the importance of good password practices.

[6] The Password Pledge was introduced by the NI Cyber Security Centre for World Password Day. It comprises the following statements: I will create strong and unique passwords using three random words. I will use different unique passwords for each of my most important online accounts: such as email, social media and banking. I will turn on a second layer of security.

Fig. 1. EXCITE Welcome Display - Individual Game Portals from the QUB campus.

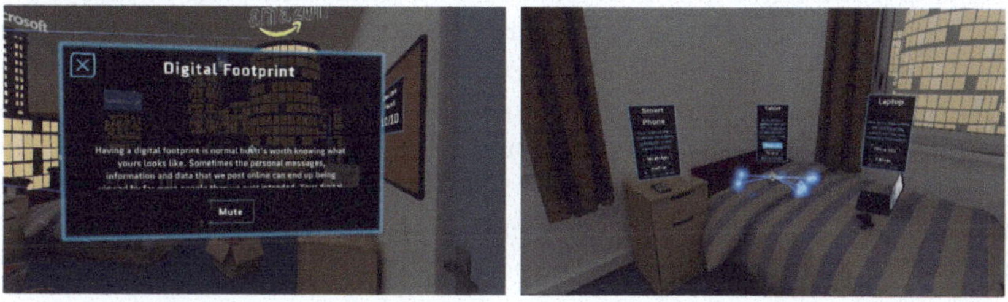

Fig. 2. EXCITE *Your Digital Footprint* Immersive Experience Example

Social Engineering. This scenario takes place in the student's room. It presents the story of an international student and a sequence of events representing common social engineering attempts through postal delivery scams, voicemail, text, and phishing email attempts. The student endeavours to identify the socially engineered contact versus benign contact in each case. The concluding message reminds the student how to avoid falling victim to a scam.

Your Digital Footprint. This scenario takes place in the student's room. The introduction highlights the digital trail created with every online interaction. In this experience, the student identifies the services that they use from the devices located around the room. As the service is selected, the connections between this service and other services/devices in the room are identified to provide an illustration of the interconnection between services and the individual digital footprint that this creates. At the end of the experience, a summary is provided as to how students should review and take ownership of their digital footprint.

Secure Remote Working. This scenario takes place in a QUB café. In this experience, the student is asked to explore the environment and identify which devices are secure for remote working/studying. There are ten devices to be identified within a set time. Secure and insecure devices are revealed with indi-

Fig. 3. EXCITE Game Completed and Progress Displays

vidual explanations. At the end of the experience, a summary of best practices for secure remote working/studying is provided.

Protecting your Data and Devices. This scenario takes place in the student's room. This experience builds on the learning from the previous experiences giving the student the context of a group work assignment and asking them to identify the data and devices in their room that they should protect physically or digitally. There are eight elements to identify (data/devices) with an explanation about their security displayed as each is discovered. The concluding message reminds the student to consider their environment, data, and devices to avoid cyber threats.

Social Media Use. This scenario takes place in the QUB library. It presents the story of a healthcare professional student on placement, their interaction on social media and the consideration of e-professionalism. The student is presented with a series of social media posts and asked to place them in a social, professional, or never post category. At the end of the experience, a summary is provided to highlight the dos and don'ts of social media use, the importance of e-professionalism and the link to professional registration.

Incident Reporting. This scenario takes place in a QUB café. The experience is designed to reinforce the importance of incident reporting and the specific reporting mechanism within QUB. The explanation of the reporting process is presented by the Director of the QUB ACE-CSE. This is followed by a whack-a-mole game to identify the cyber security incidents from benign events related to all the cyber security topics covered in the experiences. The concluding message highlights what happens after the student reports an incident.

8 Deployment

As previously detailed, two versions of the immersive experiences were developed: 1) a desktop application available for students to download and run on their own device, supporting broad accessibility of the training; and 2) a VR headset application providing the fully immersive environment for focused, small group training. In this section, we briefly outline the deployment approach for each version.

To encourage completion of the cyber security training and to avoid unnecessary data capture, the information recorded at sign-in to the application is deliberately limited. In EXCITE Phase 2, the students were asked for their views on data capture for the purpose of the training and strongly indicated that there should be very limited information requested. These views were considered in the design of the sign-in process.

VR Application: The VR headset application is offered through in-person training sessions. In these training sessions, the student can provide their name, university email address, and School/Faculty information. The email address can be used to share a certificate of completion and a cyber security information sheet containing additional information referred to during the immersive experiences. The School/Faculty information can be used to monitor the engagement with the training across disciplines. Additionally, students can be asked to optionally answer the following three questions before and after they complete the experiences:

1. How aware are you of how to protect your information online and your electronic devices? (ranked on a scale from 1–10, with 1 being not very aware to 10 being very aware)
2. How confident are you in your ability to protect your information online and your electronic devices? (ranked on a scale from 1–10, with 1 being not very confident to 10 being very confident)
3. How serious do you think the consequences would be for you if someone obtained your information online or was able to access your electronic devices? (ranked on a scale from 1–10, with 1 being not very serious to 10 being very serious)

The answers to these questions can be used to assess any change in the students' understanding and awareness of how to protect themselves and their devices online, how confident they are in their ability to protect themselves and their devices, and their awareness of the potential consequences of poor cyber security practices following the completion of the cyber security training.

To start the application, the student selects 'New User', which generates a unique identifier assigned to the student. This identifier is used to log their progress through the experiences and can be used to re-enter the experience later, if necessary. For the VR experience, the student can read out the identifier to the trainer for recording purposes. On completion of the VR training, the trainer can email the student a certificate of completion and an information sheet containing the additional information referred to during the training.

Desktop Application: The desktop application is available for students to download from a QUB hosted webpage. In practice, this is embedded in the QUB Virtual Learning Environment, Canvas, in a cyber security training module, which also includes supporting information on each of the cyber security topics, links to QUB cyber security policies and practices relevant to students, along with some quizzes for self-evaluation. Students can download and launch the application on their own device. As with the VR application, the student selects 'New User', which generates a unique identifier assigned to the student. This identifier is used to log their progress through the experiences and can be used to re-enter the experience and continue or repeat the training, if necessary. On completion of the desktop training, the student can submit a request for a certificate of completion by providing their unique identifier, university email address, and their School/Faculty. The additional information referred to during the training is directly available in the Canvas module.

9 Rollout Experience

The EXCITE application has been showcased to QUB students (and staff) at various events on campus and has been available to students since Autumn 2023. The experiences have also been demonstrated in various regional and national cyber security education, skills, and industry forums with very positive feedback. Through the EXCITE VR Roadshows (open demo/trial sessions at various locations on campus), we identified that a shorter demo VR version consisting of one or two experiences would be beneficial to increase throughput during the sessions. Subsequently, this two-experience version has also been used to showcase the project and inspire school pupils to consider studying or working in cyber security e.g., at the Northern Ireland CyberFirst EmPower Girls event and the Digital Schoolhouse Project Stranmillis Festival of Play. Although designed for third-level students, the VR is engaging, and some simple cyber security messaging can be explained to the school pupils. The demo version has also enabled quick trials by other UK universities as one experience takes less than five minutes to complete. Through our ACE-CSE School liaisons (representatives in QUB Schools responsible for supporting the cyber education strategy in those disciplines), the EXCITE desktop application has been made available in various modules (with introduction either at School induction or in a specific course). Completion has been most successful when the training has been embedded in a module or programme. For example, within our School of Electronics, Electrical Engineering and Computer Science (EEECS), the training is now delivered to students as part of their compulsory Stage 2 *Professional and Transferrable Skills* module prior to undertaking their placement year. In 2023, approximately 200 EEECS students completed the training. Similarly, in the School of Pharmacy, students are required to complete the training before attending experiential learning and in 2023, approximately 150 Pharmacy students completed the training. Completion across other Schools/programmes has been sporadic. As a result of this and based on interest from other UK universities to access the training, to increase accessibility and adoption within QUB and across UK HEIs, a web-browser based version has been developed. This enables direct access to the EXCITE application from any device.

10 Conclusion

In this chapter, we have presented the design and development of the EXCITE cyber security immersive experiences. To the best of our knowledge, this is the first cyber security immersive training designed with and for university students. Our goal with this training is to help third-level students, from across disciplines, to understand cyber security better, develop their knowledge and skills in this area, as well as enhance their understanding of the potential consequences of poor cyber security both personally and professionally.

Embracing AI in Cyber Security Education: A Postgraduate Case Study

Eckhard Pfluegel(✉), M. Rehan Usman, David Pickard, and Vesna Brujic-Okretic

Kingston University London, Thames, UK
{e.pfluegel,r.usman,d.pickard,vesna}@kingston.ac.uk

Abstract. While there is an increasing emphasis on preparing students for modern digital tools such as AI, educators face the dilemma of weighing up the advantages and disadvantages of this technology, including the problem of detecting its misuse. The authentic use of AI tools in education is an urgent requirement recently attracting significant attention. In this work, we are interested in the cyber security discipline and pedagogical approaches that may respond to the abovementioned challenge, focussing on assessment design tailored to integrating generative AI tools such as ChatGPT and the issue of their misuse in creating student assignments. We report on a case study involving a network and information security postgraduate module where students were educated about these tools and how they can help brainstorm ideas, outline reports, and fix grammatical mistakes in English writing. Apart from submitting reports, there were two additional components for their assessment: a viva session and a hands-on student demonstration. We argue that our approach can successfully assess authentic AI use. In addition, student feedback shows high satisfaction with the revised assess- ment format, the one-to-one interactions with lecturers and the role-play aspect of vivas. An analysis of assessment outcomes also indicates improved module performance. Our case study demonstrates the possibility of embracing AI in cyber security education while respecting its ethical use, with the additional benefits of enhancing student experience and academic performance.

Keywords: AI in Education · Authentic Use of AI · Cyber Security Education · Generative AI · Higher Education

1 Introduction

The rapid advancement of Artificial Intelligence (AI) technologies in recent months has fundamentally transformed the higher education landscape for virtually all subject domains, including cyber security [1]. This makes it crucial for students to develop the skills necessary to embrace AI tools, prepare themselves for rapidly changing job market expectations, and compete with peers regarding their digital skill sets [2]. A critical understanding of AI's multiple dimensions is mandatory. As educators, we must help students overcome this

P. Legg et al. (Eds.): CSE 2024, LNNS 1213, pp. 116–136, 2024.
https://doi.org/10.1007/978-3-031-77524-6_7

complex challenge by repeatedly addressing several aspects: The media's portrayal of AI often varies between sensationalism and reality, contributing to a public perception that is both fascinating and wary [3]. This can make it difficult for a neophyte to approach the technology. AI is not a new technology in computer science, but only people trained in this subject area will be familiar with technical concepts such as machine learning and neuronal networks. Students in today's cyber degree programs will have difficulties understanding these aspects. In research, AI offers new opportunities for discovery and building on existing results. Yet, it also risks becoming prone to the noise of many recent publications, which hinders students from accessing relevant, high-quality articles that critically exposition correct, up-to-date information. Hence, AI can be seen as a double-edged sword for students. It provides powerful tools for learning and problem-solving while creating dilemmas and obstacles. Educators face the critical decision of embracing AI, integrating it into the curriculum to enhance learning, or resisting its influence, maintaining traditional educational means [4].

ChatGPT in the classroom offers a novel, interactive approach to teaching and learning cyber security. ChatGPT can be an interactive chatbot for lecturer-guided workshop activities to illustrate model answers to revision exam questions or facilitate discussions between students and the lecturer [5]. This approach enhances student engagement and allows immediate clarification and deeper exploration of complex top- ics. It might also illustrate the limits of AI technology as inevitably hallucinations ([2]) - mistakes, sometimes merely in the form of minor, subtle inaccuracies only identifiable by an expert - will occur at some point. Lecturers can also use Chat- GPT to generate educational content, such as realistic scenarios for problem-based learning or instructions for workshop exercises for students to apply their knowledge and problem-solving skills. In addition to structured workshops, ChatGPT supports student-led self-exploration of cyber security knowledge through prompt engineering [6]. By crafting specific prompts, students can interact with ChatGPT to learn about various aspects of cyber security, such as ethical hacking techniques, information secu- rity frameworks, and cryptographic principles. As they refine their prompts, students can ask more precise and insightful questions, which leads to a deeper understand- ing of the subject matter [7]. This method empowers students to take control of their learning but requires patience and perseverance. The role of educators is to encourage curiosity and independence within this process. Thus, we can conclude that integrating ChatGPT for guided and self-directed activities creates a balanced and comprehensive educational approach. By adopting ChatGPT for their teaching practice, educators can create a flexible and effective learning environment where students can develop critical thinking and practical skills for future cyber security careers.

In the field of cyber security, an essential demand on the workforce is the capability to anticipate, formulate and mitigate diverse threats [8]. The ability to articulate com- plex concepts clearly and effectively is crucial to developing these skills, making essay writing a critical component of cyber security education. As generative AI (GenAI) technologies such as ChatGPT have advanced and become available to students and educators, they have given rise to novel approaches to enhancing the teaching of essay writing methodologies. These AI tools can help students brainstorm ideas, structure their drafts, refine language style and correct spelling and grammatical mistakes [9]. GenAI can also enable students to practice writing for different audiences, a skill vital for effective communication with technical and non-technical stakeholders. This can prepare cyber security students to clearly and persuasively convey complex issues. However, a fundamental problem is the potential misuse of these technologies, such as generating an entire essay with minimal individual contribution [10]. This raises concerns about cheating and the decline of educational standards. Students might be tempted to rely too heavily on AI, and detecting this is now difficult, if not impossible. Addressing these concerns requires a careful approach. Educational institutions must devise and implement robust guidelines and monitoring systems to prevent misuse while promoting the ethical use of AI as a learning tool for students.

This study was motivated by our aim to enhance student critical thinking and understanding of AI in the context of cyber education. We wish to educate students about the ethical, authentic use of AI tools in their cyber security learning path. We believe that our approach should consist of embracing rather than combatting AI; the key objective reported in this article is the successful integration of AI tools for teaching and learning in a cyber security module. This includes their use in the classroom to acquire new knowledge, demonstrate and illustrate complex concepts, and for assessment. We present a case study based on a postgraduate module run at Kingston University, a Higher Education institution in the UK, to provide deeper insights into this issue. Integrating AI, particularly ChatGPT, into Kingston Univer- sity's cyber security curriculum has positively impacted course metrics and student and lecturer experiences. Our approach to assessment enhances engagement, satis- faction, and flexibility and seems especially suitable for postgraduate students in a smaller class setting.

This paper is organised as follows: Sect. 2 is the literature review. We explain the design of our methodology in Sect. 3, followed by the presentation of our case study in Sect. 4. Section 5 is the conclusion.

2 Literature Review

Since the ChatGPT 3 web app was made available to the public in late 2022, there has been an explosion of publications on ChatGPT and, more generally, GenAI in edu- cation. In this review section, we will restrict ourselves to papers specifically relevant to this article's work: those examining the use of GenAI for assessment, particularly its ethical aspects. Note that we do not directly cover aspects of detecting the use of GenAI ([11–13]), but instead examine proposed work for collaborating with GenAI Tools in the context of learning and teaching, both as a learner or an educator. The key references on assessment and AI, reviewed in Sect. 2.1, are the papers [5,12,14–19]. To the best of our knowledge, [18] is the only educational paper with direct application to the cyber security domain. There are a few related case studies in the educational literature [7,10, 20–26] which will be discussed in Sect. 2.2.

2.1 AI for Learning and Teaching

The work in paper [5] explored the implementation of ChatGPT in university teach- ing, focusing on its benefits and potential challenges as perceived by professors at Trakia University in Bulgaria. The study found that a significant number of professors appreciated use of ChatGPT for its ability to enhance student engagement, provide personalized learning, and assist with time-consuming tasks. Specifically, 60.9% of professors agreed that ChatGPT aids in managing extensive teaching activities, while 59.8% believed it helps to engage and stimulate students. However, the study also revealed substantial concerns, with 73.6% of professors worried that students might over-rely on AI-generated content without critically assessing its accuracy, potentially compromising academic integrity and critical thinking skills. Despite these concerns, the overall perception of ChatGPT is positive, with many educators recognizing its potential to innovate educational practices and encourage critical thinking and creativ- ity. The study concluded that while ChatGPT offers significant benefits for education, it is crucial to address ethical issues and establish guidelines to prevent its misuse.

As ChatGPT is the most common AI tool used, a study conducted in [12] analyzed the effectiveness and implications of using ChatGPT in engineering education assess- ments across multiple institutions in Australia. The study revealed that ChatGPT can successfully complete some types of assessments, particularly multiple-choice quizzes and basic programming tasks, often achieving passable or even high scores with min- imal input adjustments. However, it falls short on more complex assignments that require deep understanding, critical thinking, and specific contextual or reference inte- gration. The results showed that while ChatGPT can generate accurate and relevant responses for simpler tasks, it frequently provides incorrect or incomplete answers for more demanding questions. The study concludes that to uphold academic integrity and ensure comprehensive assessment of student capabilities, educational institutions must redesign assessment practices. Incorporating AI's potential to enhance learning while developing robust mechanisms to prevent misuse is crucial for future educational strategies.

In [14], authors examined the challenges of traditional assessment methods in the context of GenAI and proposed a new framework for authentic assessments to address these issues. Their study identified that conventional assessments often fail to capture diverse student knowledge and are vulnerable to exploitation through AI tools. To counter this, the authors suggested authentic assessments, which focus on real-world applications and critical thinking, as a solution to enhance academic integrity and the learning process. The proposed model integrates GenAI tools into authentic assess- ments, aiming to reduce academic misconduct and better evaluate students' genuine understanding and skills. The study concludes that this approach not only improves student engagement and professional competencies but also safeguards against the mis- use of AI in education. The framework provides practical examples and strategies for educators to redesign assessment practices, ensuring a more accurate and meaningful evaluation of student abilities in real-world scenarios.

Similar to previous studies, [15] explores the integration of GenAI tools, specifically ChatGPT 3.5, into experiential learning activities to enhance authentic assessment. The study demonstrates that GenAI tools can significantly enrich educational expe- riences by fostering reflective thinking, practical skill development, and critical problem-solving. For instance, in a conceptual example within an operations man- agement context, students engaged in a virtual healthcare simulation enhanced by GenAI tools, which supported the experiential learning cycle stages from concrete experiences to abstract conceptualization. The study found that using GenAI tools for advanced scenario challenges and continuous improvement tasks improved students' understanding and application of Lean methodologies. Additionally, authentic assess- ment practices, such as performance metrics, feedback, and peer evaluations, were enhanced by GenAI, promoting deeper learning and practical skill application. The study concludes that while GenAI tools hold transformative potential in education, their responsible and ethical use is crucial. Future research should focus on practi- cal implementations to validate these findings and further explore GenAI's role in educational settings.

In [16], integration of GenAI tools in experiential learning activities to improve authentic assessments in higher education was investigated. The study demonstrated that GenAI tools, such as ChatGPT, can significantly enhance educational experi- ences by facilitating dynamic, interactive, and real-world applicable learning activities. The results revealed that GenAI tools effectively support the stages of the experi- ential learning cycle, including concrete experience, reflective observation, abstract conceptualization, and active experimentation. For example, AI-driven simulations and virtual reality experiences were shown to provide students with immersive, prac- tical learning scenarios that foster critical thinking and problem-solving skills. The study also highlighted that GenAI can help create personalized learning pathways and provide immediate feedback, thus promoting continuous improvement and deeper understanding. The conclusions emphasize the transformative potential of GenAI in creating authentic assessments that align closely with real-world contexts and learn- ing outcomes. However, the authors caution against over-reliance on AI, stressing the importance of maintaining academic integrity and ethical considerations. They advo- cate for the responsible use of GenAI tools, recommending continuous monitoring and evaluation to optimize their benefits while addressing potential risks. The study calls for educators to strategically incorporate GenAI into their teaching practices to enhance learning effectiveness and foster meaningful educational experiences.

The authors of [17] explored the transformative impact of GenAI on assessment design in education, focusing on the challenges and opportunities it presents. The study categorized assessments into learning-based and action-based types, emphasizing the need for innovative strategies to ensure academic integrity and effective learning out- comes. Key findings indicated that while GenAI can enhance the learning experience, it also necessitates robust design strategies to prevent misuse and ensure authenticity. The authors proposed a structured process for integrating GenAI into assessments, involving steps such as determining assessment types, applying authentic design tech- niques, validating assessments against GenAI capabilities, and clear communication of expectations. The study concludes that the ethical and effective use of GenAI in educational assessments can significantly enhance learning outcomes and prepare stu- dents for future challenges, provided that the assessments are carefully designed to balance learning, integrity, and practical application.

In [18], authors investigate the impact of ChatGPT on cyber security education assessments, focusing on designing robust assessments to ensure academic integrity. The study uses diverse cyber security questions, analyzed through various prompts tailored to different academic levels, and applies statistical tools to evaluate ChatGPT's performance. Key findings reveal that ChatGPT can generate high-quality answers for simpler questions but struggles with complex, context-specific tasks requiring critical thinking and detailed analysis. Fact-checking and the type of question significantly influence the quality of AI-generated responses, with complex, fact-checked ques- tions being more challenging for ChatGPT. The study concludes that to maintain assessment integrity, educators should design tasks that require deep understanding and critical evaluation, use recent academic references, and incorporate fact-checking mechanisms. The authors recommend that cyber security assessments be continuously updated to align with evolving technologies and emphasize the importance of ethical guidelines to mitigate the misuse of AI in academic settings.

From information retrieval and management perspective, [19] explores the effects of AI-based ChatGPT on higher education libraries, focusing on its potential to trans- form library services. The study reveals that ChatGPT can significantly enhance the efficiency of information retrieval and management, offering personalized recommen- dations and 24/7 accessibility. Key findings indicate that ChatGPT can automate routine tasks such as cataloguing and sorting, thus freeing up library staff to engage in more complex, value-added activities. However, the study also identifies challenges, including the risk of misinformation and the necessity for robust digital literacy among users to ensure ethical use. The authors conclude that while ChatGPT presents numer- ous advantages, such as increased accessibility and personalized user experiences, it is crucial to implement comprehensive training programs and ethical guidelines. This will maximize the benefits of AI integration while mitigating risks, thereby enhancing the overall educational experience and maintaining academic integrity.

The integration of GenAI tools, in the context of learning and teaching, holds signif- icant potential to transform educational practices for both learners and educators. The above studies have shown that these tools can enhance student engagement, provide personalized learning experiences, and support the development of critical thinking and practical skills through authentic assessments and experiential learning activi- ties. ChatGPT, for instance, excels in automating routine tasks, generating interactive learning scenarios, and offering real-time feedback, thereby allowing educators to focus on higher-order teaching activities. However, the use of GenAI in education must be approached with caution to ensure ethical use and maintain academic integrity. The responsible deployment of these technologies, guided by continuous evaluation and strategic integration, can lead to more effective, dynamic, and meaningful learning experiences, preparing students for real-world challenges while supporting educators in their instructional roles which is the main focus of this article.

2.2 Educational AI Case Studies

The paper [10] investigates the integration of ChatGPT in a Vietnamese educational setting based on empirical evidence from student and staff surveys and semi-structured interviews. In total, 73 undergraduate students participated in their first or second year of technical studies. One of the strong points of the paper is the critical awareness of the drawbacks of using AI, including the risk of students becoming overly dependent on AI tools, taking a more passive approach to learning and having less ability to evaluate information critically. Students also report worrying about fake news and security threats being amplified due to the new generation of AI tools. On the other hand, in contrast to our study, the paper does not investigate formal assessment incorporating AI and is restricted to reporting opinions.

A strong motivation to adapt assessment for using GenAI, as proposed in our study, is one of the outcomes of the paper [21], an empirical study analysing ChatGPTs ability to solve assignment tasks in an undergraduate Bachelor of Business Administration course in the Abu Dhabi Emirates. Using a quasi-experimental design, graded work was randomly chosen anonymously and compared with answers generated by ChatGPT. It was concluded that ChatGPT poses a threat to traditional assessment, and the need to discuss this issue with all relevant stakeholders openly was formulated. However, the study does not go beyond this insight and investigate the impact, effect or use of ChatGPT pre- or post-assessment, as is the case in our work. The need to further comprehend the students' perception of ChatGPT, as done in [10], is also mentioned.

The paper [26] adopts the wider context of GenAI (including Bing Chat) for STEM Education and examines the benefits of using these technologies for student learning. The paper focuses on prompt engineering, where the interactions with conversational AIs are analysed. It is argued that chatbots can be modelled as objects-to-think-with, a concept defined in constructionist learning theory [27]. The case study element of the paper does not refer to a specific educational programme or setting but to the AI chatbots themselves. This is also one of the limitations, as no direct student data or feedback is considered.

The prompt engineering aspect is very relevant to Section X in this article, and the methodology is similar except for our domain, which is cyber security.

Relying purely on ChatGPT as a substitution for real human teachers has dis- advantages, as established in the paper [20]. Based on a sample of 39 students in a chemistry course in the United Arab Emirates, their work evaluates student per- formance on learning outcomes categorised using Bloom's taxonomy. Using a control group, it is shown that some aspects of knowing, applying, and reasoning fell behind when relying exclusively on ChatGPT. On the other hand, student engagement with the new technology was generally very good. The study does not focus on Higher Edu- cation and is potentially less relevant for the work in our study. Also, the evaluation is short-term, specific to the curriculum chosen, and some of the conclusions will need further revisiting and refinement.

The study in [24] is a case study in a data analytics course comprised of 74 Business and ICT undergraduate and postgraduate students in Australia. The paper shows the perceived advantages and disadvantages of ChatGPT using a detailed analysis of questionnaire data. Students participated in a video tutorial and explorative session using ChatGPT, after which they completed an online questionnaire. The findings are interesting and broadly in line of similar surveys, some of the recommendations appear preliminary.

A wide range of opinions are gathered and analysed in [25], a three-stage instru- mental case study on using AI chatbots in general education. Pertinent stakeholders, including individuals sending public social media messages, are collected and pro- cessed using cluster analysis: ¿2000 tweets, interviews and feedback relating to user experience from a sample of experienced educators in a week-long study. The findings contribute to the ongoing debate on this topic. Ethical issues are raised, the need for further action to transform education is illustrated, and an interesting aspect is the inclusion of emotional analysis in the data. On the other hand, in contrast to ours, the study is only quantitative, and the case study subject is not a specific educational setting.

We also report on three papers with case studies involving the use of Chat-GPT in a computer science context and programming.

In [23], a workflow explains how to create lab assignments for a bachelor's and post- graduate computer science module involving the creation of Python code, including using sophisticated image recognition libraries. It is concluded that ChatGPT is very helpful in creating skeleton instructions and can be refined to include model answers, but it requires careful vetting to validate its output. Otherwise, the created code might not be optimal, and included references were often found to be artificially fabricated.

The initial study [22] reports on the impact of ChatGPT on learning in a pro- gramming course for first-year students at a Japanese university. Using a pre-and post-questionnaire integrated into an eight-week duration, students used ChatGPT for their programming assignments. The levels of student interaction with the AI tool were analysed and found to correlate with the intellectual challenge levels of the vari- ous assignment milestones. The article concludes with an open note on the need for a detailed analysis of student feedback.

Another research paper on learning introductory Python programming with the help of Large Language Models (LLMs) is [7]. The paper explores using a custom learning platform that employs ChatGPT to create personalised learning materials for high school students. This method enhances traditional computer science education, and the case study reveals an increased student engagement and a shift in instructors' roles from conventional teaching to mentoring. Their conclusion also indicates the potential of LLMs for personalised learning to a global audience in other fields.

3 Methodology

This section describes our methodology for integrating ChatGPT in teaching and learning. This will include specific examples of activities undertaken in the classroom and experiences gained from a lecturer's perspective. Regarding integrating AI tools into classroom activities, we first describe prompt engineering as an interactive feature that can allow students to explore complex concepts in cyber security independently, but we also note some pitfalls when using AI for student-led learning. We then report on an experiment where lecturers used ChatGPT to create workshop instructions for exercises, including a critical review of the out- comes. Finally, we investigate how students can be taught the ethical, authentic use of ChatGPT for assignment writing. This is a central theme of our work and prepares the assessment design, discussed in the next section.

Table 1 Summary of published Case Studies: main features, educational domain, evaluative contribution and consideration of assessment issues - Table to be inserted

3.1 Interactive Prompt Engineering

Prompt engineering denotes the continued, interactive use of an AI chatbot in a session, building on provided output to craft new queries and refining AI-created content and structure. In this section, we document a simple prompt engineering ses- sion, starting with a query of explaining basic cyber security terminology, followed by a refinement request to shorten the output, and finally, an instruction to obtain a narrative based on given keywords as a final clarification query.

A typical dialogue with ChatGPT might be initiated with a request for knowledge (Fig. 1).

A student's first reaction might be to be slightly overwhelmed (the screenshot only shows parts of ChatGPT's response; much more information was included in the answer), especially as only a brief overview was requested. The response is given in the form of an outline, with one level of indentation and headings starting with the topics of risk assessment, security policy development, and implementation of security controls. The first topic is broken down well by differentiating between assets, threats and vulnerabilities. However, it could be made clearer that vulnerabilities are weaknesses of assets (rather than generically of systems). The second topic is potentially confusing: security policies are a form of security control; for more clarity, this could be included in the third topic. On the other hand, describing controls in terms of their preventative, detective, or corrective nature is accurate and corresponds to common practice in the literature. The topics continue with security training and awareness; these are examples of controls. These examples should be added to the list of preventative controls contained in the previous topic.

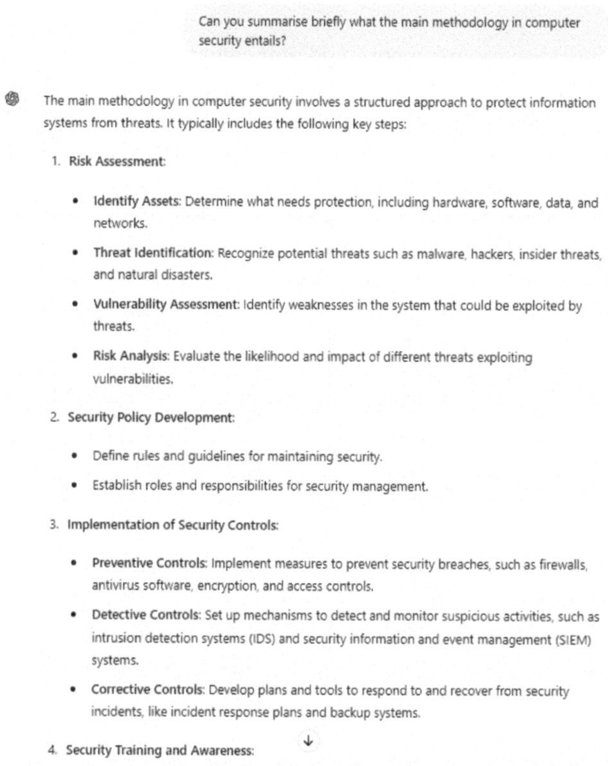

Fig. 1. ChatGPT's answer for a request for knowledge: Basic security terminology.

In summary, ChatGPT's answer contains useful and correct explanations of the theory, including useful examples. However, greater clarity could be achieved, provid- ing more nuanced knowledge to the student. A student with no prior knowledge of cyber security might not be able to recognise the full picture of these quite subtle points. On the other hand, a cyber security expert would notice them upon careful reading. Next, the student could use prompt engineering to achieve a shorter response, as shown in Fig. 2.

The obtained summary falls short of the 50 words by containing only 37 words. Being very condensed, the important notions of asset and vulnerability

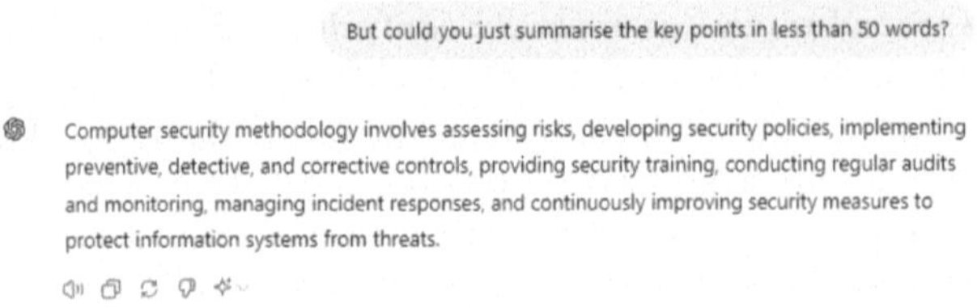

Fig. 2. Use of prompt engineering to shorten the previous response.

are omitted, and threats are mentioned somewhat out of context at the end of the paragraph. There appears to be a focus on assessing risk. Amongst the suggested risk mitigation tech- niques, the answer given by ChatGPT contains a mix of terminology (preventative, detective and corrective controls) and examples: security training is usually preventative, audit and monitoring is detective and instance response is corrective. However, this might not immediately be clear to a student, as it is not directly mentioned. The point suggested last, to continuously improve security measures, is a principle for security management: periodic review, which could have been emphasised more, potentially in a separate, self-contained sentence.

After evaluating the outcome of the second query instruction, we can confirm that ChatGPT has been successfully used to condense the output. However, better using up the given word count could have provided more useful information. The same problem as in the previous response, relating to a certain fuzziness in the provided content, which is potentially difficult to recognise by a novice in the field, persists.

Another approach uses ChatGPT to provide a narrative based on given keywords. For example, after issuing several queries, a student could collect key concepts in the various answers and then seek clarification, as illustrated in Fig. 3.

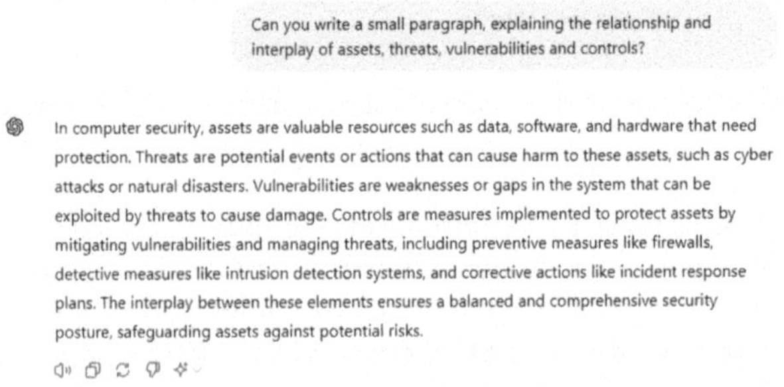

Fig. 3. Creation of narrative, based on given key concepts.

The resulting answer is of good quality regarding the created text's style, language and structure, but it could be improved in several ways. To illustrate the key concepts, ChatGPT provided some examples: the asset categories of data, software, and hard- ware are given, although it would be important to mention the crucial people category in addition. The examples given for three different types of controls (preventative or proactive, retrospective or reactive, and corrective) are accurate and correspond to popular examples presented to students in an educational context. Threats are char- acterised as potential events or actions; at this point, it would be useful to refer to the notion of attack as a threat event or implementation. Vulnerabilities are described correctly, although the terminology mitigation would be more suitable when talking about the concept of risk, which coincidentally was not provided as a keyword in the question prompt but would have perhaps been important to elaborate on. Finally, the statement that the interplay between these notions (referred to as elements) ensures a balanced and comprehensive security posture is questionable - the interplay merely describes typical scenarios in cyber security, and managing the security of such a system is successful, if a suitable security assessment and management framework overlooks the various actions of the participants in this strategic, interactive scenario.

In summary, the provided paragraph is certainly not incorrect. Still, as in the previous examples, students can make misleading interpretations due to the lack of completeness of the provided information or ChatGPT's specific focus. Compared to previous works, as discussed more in detail in the literature review section, our findings on prompt engineering show that it would be insufficient to rely solely on ChatGPT as a teacher, confirming the central message in [20]. On the other hand, we have not systematically evaluated this approach in contrast to the cited paper.

3.2 AI-Generated Workshop Exercises

This section will discuss the ability of AI tools to generate teaching resources. The emphasis is on instructions for workshop exercises, and the advantages and disadvantages of this approach are highlighted.

Many authors have noted ChatGPT's remarkable functionality for creating teacher resources, which has been applied to numerous specialist technical disciplines, includ- ing programming and cyber security. We trialled this approach for security and risk assessment workshops and found that ChatGPT significantly sped up the creation of well-written and well-structured materials. Particularly helpful was dividing the instructions into clearly defined sections with suitable headings. Creating a formatted Word document from the AI-generated text was not difficult. It was also possible to create LaTeX documents in different experiments, including the suitable typesetting of mathematical equations.

Figure 4 shows a simple exercise on security economics created by ChatGPT. It was remarkable that the lecturer had intended to introduce this new topic to the students for several years but, due to the ongoing process during term time, never managed to focus enough. This new AI tool made creating a seed for some slides on this topic easy, followed by a seed for the exercise. Some minor editing was required to bring it up to the desired standards. Students reported their satisfaction with the lecture and workshop that were delivered. The lecturer was open about the fact that ChatGPT had been used and demonstrated the use of prompt engineering for related tasks in class.

Exercise on Using ROI, SLE, And ALE for Security Economics

Objective:

To demonstrate how to use ROI, SLE, and ALE to make informed decisions about security investments.

Scenario:

You are the IT security manager of a small company that processes sensitive customer data. The company has recently experienced a security breach that resulted in the loss of 500 customer records. The company's CEO has asked you to provide a report on the potential financial impact of the breach and to recommend measures to prevent future incidents.

Tasks:

1. Calculate the SLE.
2. Estimate the ARO.
3. Calculate the ALE.
4. Calculate the ROI of implementing security measures.
5. Recommend security measures.

Fig. 4. Exercise created by ChatGPT.

On the other hand, an experience with a different topic showed that the limits of this approach, as a subtle hallucination in the instructions for applying Bayesian statistics to risk analysis of a fictitious cyber security scenario, was difficult to spot initially. This created confusion and time delays for the lecturer, who had little expe- rience in this area. The issue was the generic use of the term control by ChatGPT in a context where it only made sense as a reactive control, for example, an IDS, as the exercise involved the consideration of the false-positive rate. The exercise text, detailed in Appendix A, contained the sentence probability of the control preventing a cyber attack given that none was attempted, also known as the false negative rate" which is highly confusing, especially to a student unfamiliar with this topic. In a later iteration, the issue became clear after inspecting other examples on the Internet. An IDS and the description as a false positive rate replaced the generic control and false negative terminology.

In conclusion, the more expert knowledge a ChatGPT user brings into the session, the better they will be able to avoid time-consuming problem-solving, as reported above.

3.3 Assignment Writing

In this section, we will discuss using ChatGPT for writing assignments. We will focus on using AI-created input as the first version of a document by creating the concept of a seed.

Definition 1. *Suppose the text output of an AI tool is used as a first draft of a document for further editing by a human author. We will refer to this text as a seed for the final document. This concept may apply to content and structure separately or simultaneously. Hence, we may also differentiate between content seed and structural seed.*

For example, a structural seed could be an outline for an assignment structure, a paper's table of contents, or a plan for a PowerPoint presentation. A typical content seed is a single paragraph serving as a first draft, which will then be revised, edited and expanded to yield the final version.

We will illustrate this idea by revisiting the narrative explaining basic security ter- minology presented in Sect. 3.1, and by addressing the suggestions for improvement of the write-up given in the same section. Figure 5 recalls the initial seed.

In computer security, assets are valuable resources such as data, software, and hardware that need protection. Threats are potential events or actions that can cause harm to these assets, such as cyber attacks or natural disasters. Vulnerabilities are weaknesses or gaps in the system that can be exploited by threats to cause damage. Controls are measures implemented to protect assets by mitigating vulnerabilities and managing threats, including preventive measures like firewalls, detective measures like intrusion detection systems, and corrective actions like incident response plans. The interplay between these elements ensures a balanced and comprehensive security posture, safeguarding assets against potential risks.

Fig. 5. Seed paragraph, provided by ChatGPT.

Figure 6 shows the final version, improving the initial seed.

In computer security, *assets* are valuable resources that need protection. They typically belong to the categories of *data*, *software*, *hardware* and *people*; an entire system can be seen as a *logical* asset. *Threats* are circumstances that may lead to events such as cyberattacks or natural disasters, causing harm to assets. *Vulnerabilities* are weaknesses or gaps in the system that can be exploited by threats to result in the *risk* of an *impact* such as damage or loss. *Controls* are countermeasures designed to protect assets by closing vulnerabilities, blocking or mitigating threats, or both. This can include *preventive* measures such as access control (firewalls), *detective* measures like intrusion detection systems (IDSs), and *corrective* actions such as business continuity or incident response plans. Managing the security of a system or scenario requires the continuous, comprehensive assessment of these entities and their mutual security interactions, to minimise risk.

Fig. 6. Final paragraph.

As shown in Fig. 7, the difference between the two versions as a Word track changes document illustrates significant editing of the original seed; content has been modified, removed and added. This editing process adds value to the original AI-created content, and this is what students need to be trained in. The initial seed can usually be obtained quickly, but the subsequent changes can be slow and difficult. The more context information and knowledge are available to the student, the easier the process will be, but it will still require considerable expertise. From this, it becomes clear that AI cannot replace writing skills [28]. The topic of using AI tools for writing essays is complex, and we cannot explore it exhaustively in this section. Other aspects are covered in the literature, for example, the series of articles on AI and English as a foreign language.

In computer security, *assets* are valuable resources that need protection. They typically ~~such~~ belong to the categories of~~as~~ *data*, *software*, ~~*and*~~ *hardware* and *people*; an entire system can be seen as a *logical* asset~~that need protection~~. *Threats* are circumstances ~~potential events or actions~~ that ~~can~~ may lead to events such as cyberattacks or natural disasters, causinge~~ge~~ harm to ~~these~~ assets~~, such as cyber attacks or natural disasters~~. *Vulnerabilities* are weaknesses or gaps in the system that can be exploited by threats to result in the *risk* of an *impact* ~~cause~~ such as damage or loss. *Controls* are countermeasures ~~implemented~~ designed to protect assets by ~~mitigating~~ closing vulnerabilities, blocking or mitigating ~~and managing~~ threats, or both. This can ~~,~~ includ~~ing~~e *preventive* measures ~~like~~ such as access control (firewalls), *detective* measures like intrusion detection systems (IDSs), and *corrective* actions ~~like~~ such as business continuity or incident response plans. Managing ~~T~~the ~~interplay~~ security of a system or scenario requires the ~~between~~continuous, comprehensive assessment of these entities and their mutual security interactions, to minimise risk.~~elements ensures a balanced and comprehensive security posture, safeguarding assets against potential risks.~~

Fig. 7. Word track changes.

4 Case Study

This section presents our case study on adapting assessment for the authentic use of AI tools.

4.1 Introduction

In 2023, due to the rapid pace of change in educational technologies brought by the new domain of AI, in particular GenAI, Kingston University's approach to cyber secu- rity education adapted to embrace rather than combat AI technologies. This applied particularly to assessment, although detailed guidelines remained under development, and lecturers needed to investigate for themselves, more specifically, how to suitably adapt learning and teaching practices for their modules.

Since January 2024, this methodology has informed teaching and assessment on a Network and Information Security module, part of a Network and Information Secu- rity Master's. This course, certified by the National Cyber Security Centre (NCSC), is led and taught by Kingston University's Academic Centre of Excellence in Cyber Security Education (ACE-CSE) members who seek to enhance student engagement, critical thinking and employability for cyber security roles while ensuring the eth- ical and authentic use of AI tools in cyber security education. The curriculum for the NIS Master's program was updated to better incorporate AI into teaching and assessment. Specifically, within the module Network and Information Security, the module's updated methodology included generative AI tools like ChatGPT in both classroom settings and assignments. Assessments were adapted accordingly to ensure the authentic use of AI.

4.2 Assessment History

Before the integration of AI, the assessment methodology for this module had involved a combination of exams and coursework until the 2021/22 academic year. In 2022/23, the module transitioned to a coursework-only assessment model to better evalu- ate practical learning outcomes. This included two components: Coursework A and Coursework B. The first assessment required students to conduct an information secu- rity assessment of a fictitious scenario using methodologies such as OCTAVE [29] or NIST, resulting in a comprehensive security assessment report. An essential aspect of the grading criteria was whether this report was specific and realistic. It should mimic an industrial report written by a security consultant or an employee with the neces- sary knowledge, and it should contain technical references to specific assets belonging to the organisation, chosen from the standard data, software, hardware, and people categories. Coursework B tasked students with performing a practical ethical hacking activity in a controlled environment and documenting their results with supporting evidence. Students could conduct their activities in a simulated network using virtu- alisation or in the university's specialist cyber lab. To provide the required evidence, screenshots should be submitted with specific information, clearly demonstrating the customised environment and the software tools used, including the output results. For example, students could use Kali Linux in a virtual network running on Oracle Virtual Box, running scripted pen testing or attack activities on Metasploit.

4.3 Adapting Assessment

Responding to the rapid advancement of AI, in January 2024, the teaching team introduced two additional interactive assessment components: a 20-minute one-to-one viva conducted online and a hands-on student demonstration of their use of ethical hacking tools.

- For Coursework A, students must now participate in a viva, simulating a report hand-over to senior staff within a company scenario tolerating ethical AI use. This role-playing exercise, on the one hand, provides the opportunity to ask questions about the students' AI-enabled methodology. Still, it also enhances the realistic setting and applicability to future employment scenarios of the assessment.
- For Coursework B, the assessment now includes a one-to-one demonstration of the undertaking ethical hacking activities and, simultaneously, reduces the emphasis on the written report. During the demonstration session, which was one-to-one, the instructor had the opportunity to ask questions about the student's understanding of the underlying theory and to observe their use of ethical hacking tools. Furthermore, students were asked to reflect on their use of AI to help them create instructions for testing and attack simulation or report structuring and drafting.

4.4 Evaluation

This section evaluates our approach, which resulted in several positive outcomes.

First, we argue that our methodology can successfully distinguish between genuine and non-authentic student use of AI and addresses ethical education concerns. During the vivas and demos, conversations can be led to explore the students' understanding of the use of AI and to identify its potential misuse. Second, student feedback indicated high satisfaction with the revised assessment format. Students were pleased to interact directly with the lecturer, which otherwise would not easily be the case for everyone in such a medium-sized module. Some students were keen on the role-play aspect of the first coursework viva, which gave them insights into future professional scenarios and improved their employability skills. Lastly, an analysis of assessment outcomes reveals a significantly improved module performance compared to previous years. Although this outcome was not necessarily expected, it suggests that our approach cultivates strengths that many students seem to have and that were previously uncatered for.

5 Conclusion

We conclude this study by briefly summarising our results and discussing advantages, disadvantages, and limitations. Finally, we outline some ideas for further work.

In summary, our case study at Kingston University demonstrates the possibility of embracing AI in cyber security education while respecting its ethical use, with the additional benefits of enhancing student experience and academic performance. We feel that we have accomplished our objective: our methodology may successfully distinguish between genuine and non-authentic student use of AI.

Although numerous online resources and institutional events have been created since early 2023, self-study and hands-on experimentation are the most effective ways to become familiar with tools such as ChatGPT and Gemini. Lecturers and instruc- tors who engage with AI also inspire students to be confident in approaching these new tools. This increases their awareness and ability to potentially recognise content created by generative AI, which is an important and useful skill for any educator. The requirement for engagement has been made easier by educational institutions that mostly have recognised this need and provide access to the latest AI technologies, such as ChatGPT 4, for both students and academics. However, it is also important to reflect on the lecturer's perspective of adapting to the changes occurring due to the growing impact of AI tools in education. Not all academics are always willing or able to adopt these tools, perhaps due to feeling overwhelmed, generally uninterested, or too busy, leading to a divide among academics. A drawback could be the additional overhead for the lecturer, although this can be mitigated. Setting up additional viva slots requires time, but it can be managed efficiently using modern videoconferencing solutions such as MS Teams, and the meetings inform the marking process well.

We plan to adopt similar approaches for other cyber modules. We must address potential challenges when transitioning to the new methodologies, but we believe they can be overcome. We will also design a new module to deepen the students' technical understanding of AI and Cyber and further integrate AI aspects into the cyber cur- riculum. We recommend that other institutions consider similar adaptations in their teaching methodologies. We feel that maintaining an open-minded approach is crucial to benefit from continuous advancement in AI, both in terms of the quality of the teaching we provide and our own proficiency and continuous professional development.

Acknowledgements. We believe that any presentation of individual work - whether a research article, presentation or classroom assignment - should acknowledge the involvement of AI tools in its creation. Following this suggested practice, we acknowledge that GenAI tools were used as a seed for this paper. ResearchRabbit [30] was employed as a preliminary tool to identify research strands in the broad field of AI applications within cyber security education.

Appendix

A Workshop Exercise Created by ChatGPT

Bayesian Risk Analysis Example

Scenario

Suppose a company has implemented a new security control to prevent unauthorized access to its network. Before the implementation of the control, the probability of a successful cyber attack was estimated to be 0.05. After the implementation of the control, the company conducted an evaluation and found that the control had prevented 80% of attempted cyber attacks. What is the probability of a successful cyber attack given that the control was implemented?

Solution

Using Bayes' theorem, we can calculate this probability as follows:

- Let A = event that a cyber attack is successful
- Let B = event that the security control is implemented
- Let P(A) = 0.05 (prior probability of a successful cyber attack)
- Let P(B|A) = 0.8 (probability of the control preventing a cyber attack given that one was attempted, also known as the detection rate)
- Let P(B|A') = 0.2 (probability of the control preventing a cyber attack given that none was attempted, also known as the false negative rate)

We want to find P(A|B), the probability of a successful cyber attack given that the control was implemented.

Using Bayes' theorem:

P(A|B) = P(B|A) * P(A) / P(B)

where P(B) = P(B|A) * P(A) + P(B|A') * P(A'), and substituting in the values, we get:

P(B) = 0.8 * 0.05 + 0.2 * 0.95 = 0.23

P(A|B) = 0.8 * 0.05 / 0.23 = 0.17

So, the probability of a successful cyber attack given that the control was implemented is 0.17.

Fig. 8. Workshop exercise created by ChatGPT. The terminology for the false negative rate is wrong and should be false positive. It is also confusing as it only make sense for reactive controls such as an IDS.

References

1. Mohammed, S.P., Hossain, G.: ChatGPT in education, healthcare, and cybersecurity: opportunities and challenges. In: 2024 IEEE 14th Annual Computing and Communication Workshop and Conference (CCWC). IEEE (2024). https://ieeexplore.ieee.org/document/10427923/
2. Dempere, J., Modugu, K., Hesham, A., Ramasamy, L.K.: The impact of ChatGPT on higher education. Front. Educ. **8**, 1–13 (2023)
3. Li, L., Ma, Z., Fan, L., Lee, S., Yu, H., Hemphill, L.: ChatGPT in education: a discourse analysis of worries and concerns on social media. Educ. Inf. Technol. **29**(9), 10729–10762 (2023)

4. Kadaruddin, K.: Empowering education through generative AI: innovative instructional strategies for tomorrow's learners. IJBLE **4**(2), 618–625 (2023)
5. Kiryakova, G., Angelova, N.: ChatGPT-a challenging tool for the university professors in their teaching practice. Educ. Sci. (Basel) **13**(10), 1056 (2023)
6. Dai, Y., Liu, A., Lim, C.P.: Reconceptualizing ChatGPT and generative AI as a student-driven innovation in higher education. Procedia CIRP **119**, 84–90 (2023)
7. Abolnejadian, M., Alipour, S., Taeb, K.: Leveraging ChatGPT for adap- tive learning through personalized prompt-based instruction: a CS1 edu- cation case study. In: Extended Abstracts of the CHI Conference on Human Factors in Computing Systems. ACM, New York, NY, USA (2024). http://dx.doi.org/10.1145/3613905.3637148
8. Infosec Institute: Why soft skills are just as important as technical skills in cybersecurity. https://www.infosecinstitute.com/newsroom/check-out-infosecs-new-cybersecurity-soft-skills-learning-path/. Accessed 09 Jul 2024 (2023)
9. Al-Alami, S.E.: EFL learners' attitudes towards utilizing ChatGPT for acquiring writing skills in higher education: a case study of computing students. J. Lang. Teach. Res. **15**(4), 1029–1038 (2024)
10. Nguyen, T.N.T., Van Lai, N., Nguyen, Q.T.: Artificial intelligence (AI) in education: a case study on ChatGPT's influence on student learning behaviors. Educ. Proc. Int. J. **13**(2), 105–121 (2024)
11. Farrelly, T., Baker, N.: Generative artificial intelligence: implications and considerations for higher education practice. Educ. Sci. **13**(11), 1109 (2023)
12. Nikolic, S., et al.: ChatGPT versus engineering education assessment: a multidisciplinary and multi-institutional benchmarking and analysis of this generative artificial intelligence tool to investigate assessment integrity. Eur. J. Eng. Educ. **48**(4), 559–614 (2023)
13. Orenstrakh, M.S., Karnalim, O., Suarez, C.A., Liut, M.: Detecting LLM-generated text in computing education: a comparative study for ChatGPT cases (2023). arXiv. org
14. Matheis, P., John, J.J.: Reframing assessments: designing authentic assessments in the age of generative AI. In: Academic Integrity in the Age of Artificial Intelligence, pp. 139–161. IGI Global (2024). https://www.igi-global.com/chapter/reframing-assessments/339223
15. Salinas-Navarro, D.E., Vilalta-Perdomo, E., Michel-Villarreal, R., Montesinos, L.: Designing experiential learning activities with generative artificial intelligence tools for authentic assessment. Interact. Technol. Smart Educ. ahead-of- print(ahead-of-print) **21**(4), 708–734 (2024)
16. Salinas-Navarro, D.E., Vilalta-Perdomo, E., Michel-Villarreal, R., Montesinos, L.: Using generative artificial intelligence tools to explain and enhance experiential learning for authentic assessment. Educ. Sci. (Basel) **14**(1), 83 (2024)
17. Kadel, R., Mishra, B.K., Shailendra, S., Abid, S., Rani, M., Mahato, S.P.: Crafting tomorrow's evaluations: assessment design strategies in the era of generative AI. Techrxiv (2024) [cs.CY]
18. Plapp, A., Wu, J., Pan, L., Chen, C., Chua, C., Zhang, J.: Strengthening cyber security education: designing robust assessments for ChatGPT-generated answers. In: Machine Learning for Cyber Security. Lecture notes in computer science, pp. 17–27. Springer, Singapore (2024). http://dx.doi.org/10.1007/978-981-97-2458-12
19. Aithal, S., Aithal, P.S.: Effects of AI-based ChatGPT on higher education libraries. Int. J. Manage. Techn. Soc. Sci. **8**(2) 95–108 (2023)

20. Ghazali, S.A.: Department of computer science and software engineering, college of information technology, UAEU, UAE: exploring the potential of Chat- GPT as a substitute teacher: a case study. Int. J. Inf. Educ. Technol. **14**(2), 271–278 (2024)
21. Chaudhry, I.S., Sarwary, S.A.M., El Refae, G.A., Chabchoub, H.: Time to revisit existing student's performance evaluation approach in higher education sector in a new era of ChatGPT - a case study. Cogent Educ. **10**(1), 2210461 (2023)
22. Ma, B., Chen, L., Konomi, S.: Enhancing programming education with ChatGPT: a case study on student perceptions and interactions in a python course. arXiv [cs.CY] (2024) [cs.CY]
23. Pawelczak, D., Neve, A.: Using ChatGPT for creating computer science exercises - a case study from the real world. In: 10th International Conference on Higher Education Advances (HEAd'24). Universitat Polit'ecnica de Val'encia, Valencia (2024). http://dx.doi.org/10.4995/head24.2024.17366
24. Sandu, R., Gide, E., Elkhodr, M.: The role and impact of ChatGPT in educational practices: insights from an Australian higher education case study. Discov. Educ. **3**(1) (2024)
25. Tlili, A., et al.: What if the devil is my guardian angel: ChatGPT as a case study of using chatbots in education. Smart Learn. Environ. **10**(1), 1–24 (2023)
26. Vasconcelos, M.A.R., Santos, R.P.: Enhancing STEM learning with ChatGPT and Bing Chat as objects to think with: a case study. Eurasia J. Math. Sci. Technol. Educ. **19**(7), 2296 (2023)
27. Malraison, P.J., Papert, S.: Mindstorms: children, computers, and powerful ideas. Two-year Coll. Math. J. **12**(4), 285 (1981)
28. R., H.B.I., Laiya, R.E., Sarumaha, M.S.: Navigating English writing proficiency tests in the era of artificial intelligence. J. Engl. Educ. Teach. **7**(3), 480–498 (2023)
29. Alberts, C., Dorofee, A.: Managing Information Security Risks. Addison-Wesley Educational, Boston, MA (2002)
30. Cole, V., Boutet, M.: Research Rabbit (product review). J. Can. Health Libr. Assoc. **44**(2), 43–47 (2023)

A Cyber Peers Framework for Primary Schools

Jacques Ophoff[1,2(✉)] and Suzanne Prior[1]

[1] Abertay University, Dundee, UK
{j.ophoff,s.prior}@abertay.ac.uk
[2] University of Cape Town, Cape Town, South Africa

Abstract. Children are exposed to cyber threats at increasingly younger ages, necessitating early education on cyber security concepts. This has led to the inclusion of cyber security education within the UK curriculum. Primary schools play a leading role in delivering this content. However, implementing cybersecurity education at scale in primary schools poses challenges due to a combination of factors such as teacher knowledge gaps, resource limitations, and children's diverse learning needs. In addition, teachers may struggle with confidence in their cyber security knowledge and perceive that children are more knowledgeable in this domain. To better understand the challenges and potential solutions, we conducted an action research project in a Scottish primary school. The project involved 23 children aged 10–12 years, parents, class teachers, and younger children within the school. Lessons on cyber security were delivered and upon completion children worked in small groups for several weeks to design and deliver short lessons on cyber security concepts to younger year groups within the school. Based on our findings a 'Cyber Peers' framework is proposed. Central to this framework is a target group of children who can engage others within the school. They are educated on relevant cyber security concepts with pedagogical input from teachers, trusted resources, and support from cyber security experts (e.g. a digital lead). Afterwards, the target group of children will engage other classes within the school. Key considerations include ensuring the target group of children possess sufficient knowledge and confidence to engage peers, while teachers should focus on class engagement and motivation during peer-to-peer lessons. By leveraging expertise, peer-to-peer learning, and trusted resources, this framework aims to empower children to become advocates for cyber security within their schools and communities.

1 Introduction

There is a growing recognition in cyber security education of the need to provide relevant and practical instruction to children. This objective is pursued through two primary channels: parental education and the formal education system. However, both approaches present distinct challenges. While parents often feel responsible for providing cyber security education to their children, they report needing more confidence and knowledge to do so effectively (Prior & Renaud, 2024). Similarly, within the education system, despite cyber security featuring in the curriculum (Lamond et al., 2022), teachers may also struggle with confidence in their cyber security knowledge (Childers et al., 2023) and perceive that children are more knowledgeable in this domain. This could lead children

P. Legg et al. (Eds.): CSE 2024, LNNS 1213, pp. 137–152, 2024.
https://doi.org/10.1007/978-3-031-77524-6_8

to believe that they possess superior knowledge compared to the adults responsible for their education in this area (Turner et al., 2022).

Within the curriculum for England, Wales, and Scotland, there is a disparity in the cyber security topics and the ages at which they are introduced. For example, the age at which phishing is introduced is between the ages of 7–11 years in England and Wales (UK Council for Internet Safety, 2020), but not until 12–15 years in Scotland (Education Scotland, 2017). This disparity is concerning when research indicates that over a quarter of children aged 11–15 years have been a victim of a phishing attack (Kaspersky, 2023).

Given these challenges, this paper explores the potential of utilizing peer mentors to deliver cyber security education in a Scottish primary school setting. Based on the assumption that supporting resources are available, we examine the potential to utilize children's technological knowledge to scale educational initiatives within primary schools. The primary aim is to investigate whether this approach would be perceived as a viable and feasible method for addressing the existing gaps in cyber security education for children.

2 Background

2.1 Cyber Security vs. Cyber Safety vs. Cyber Resilience

Several terms are commonly used to refer to the target of security education. To clarify subsequent discussion, we provide brief definitions of three such terms. The most frequently encountered, cyber security, can be defined as *"the organization and collection of resources, processes, and structures used to protect cyberspace and cyberspace-enabled systems from occurrences that misalign de jure from de facto property rights"* (Craigen, Diakun-Thibault, and Purse, 2014, p.16). Here the focus is on reducing the risk of cyber attack with a focus on information and systems. In comparison, the focus is shifted more towards humans when referring to cyber safety, which can be defined as *"the safe and responsible use of information and communication technologies"* (Grey, 2011, p.77). This shifts the focus towards protecting users from harmful online content, conduct, and contact as opposed to the cyber security focus on confidentiality, integrity, and availability of information and systems (Renaud & Prior, 2021). Finally, the more recent notion of cyber resilience moves beyond preventative measures to include consideration of responding to, and recovering from, cyber attacks. This is more commonly linked to operational resilience and business continuity (UK Government, 2024) but can also apply at a personal level.

2.2 Cyber Security in the UK Curriculum

Formal education in the UK begins from the age of four, with primary education continuing until children are 12 years old. While cyber security features in some form from the earliest stages in the UK curriculum this varies across the countries within the UK, which raises concerns that the learning outcomes expected in cyber security at different ages may not align with children's developmental abilities, particularly surrounding password practices, a topic that dominates the curriculum in the earlier years of primary school (Lamond et al., 2022).

In turn, children have reported feeling that the cyber security concepts taught in schools do not cover what they would like to know (Turner et al., 2022). This is supported by research suggesting that teachers may not be aware of the actions their pupils are currently engaged with; for example, teachers stated they did not believe privacy would be a concern for 11–12-year-olds as they would not be posting online (Kumar et al., 2019). However, this contrasts with findings from Ofcom (2022) that indicates children are posting on social media accounts from the age of eight. This suggests that teachers may not be best placed to determine what cyber security topics should be taught to particular age groups.

Effective teaching requires that in addition to having skills in communication and empathy, a teacher needs an understanding of the topic being taught (Lipka & Brinthaupt, 1999). This is particularly true in specialized topics such as cyber security. Childers et al. (2023) investigated the impact of training teachers to deliver cyber security lessons. They found that before training teachers rated their confidence in developing and delivering lessons negatively. This increased after a week-long training course. However, it is unlikely that most teachers would be able to devote such time to this topic. Insufficient experience in pedagogy in specialized areas can significantly diminish a teacher's confidence and effectiveness in teaching the subject (Tondeur et al., 2017).

2.3 Children's Cyber Security Concerns

Another stakeholder in a child's education is the parent who often feels that the responsibility for educating their child lies with them (Prior & Renaud, 2024). With cyber safety featuring heavily in the news (Rahman-Jones & Vallance, 2023) it is perhaps not surprising that parents' concerns regarding their children online tend to be focused on cyber safety rather than cyber security (Muir & Joinson, 2020; Renaud & Prior, 2021). This is despite the real cyber security risks facing children online (Kaspersky, 2023).

Parents' concerns do not always match with the concerns of the child themselves; for example, when discussing IoT devices in the home, Turner et al. (2022) found that children had concerns about the physical security of IoT devices, a concern not shared by their parents. Turner also noted that some children felt they had a more in-depth knowledge of cyber security than their parents.

Parents or other trusted adults in a child's life need to be aware of what that child is doing online to give appropriate cyber security advice. However, these adults are not always aware of what is happening in a child's digital life (Blackwell et al., 2016). Vygotsky's (1978) theory of social constructivism highlights that a child learns best in the company of another person who has greater knowledge on the topic. However, this does not require that the other person is an adult, and it has been suggested that teenagers would welcome the use of peer mentors in this area (Nicholson et al., 2021).

2.4 Peer Mentors

The use of peer mentors in education is not a new concept and has often focused on the peer mentoring of new teachers or those studying to become teachers (Nicholson et al., 2018). Peer mentoring by pupils is common in behavioral support and positive role modeling, such as through the Big Brother Big Sisters program in the USA (Karcher,

2013). This program typically includes a significant age gap between mentor and mentee (with the mentor being well into adolescence), but in recent years there has been evidence of mentors becoming younger and the age gap decreasing.

Mentoring has been used in other subjects, such as mathematics, within the primary school age group (Thurston et al., 2020; Witt, 2020). However, most mentoring projects have followed a structured syllabus with the activities planned by a teacher in advance. To meet the need for cybersecurity lessons to concentrate on topics deemed important by the end user, this study considers the involvement of peer mentors in the development and delivery of cybersecurity education activities to their younger peers.

To date the use mentors within cybersecurity has not been explored widely. Thus, this study explores the following research questions:

RQ1: How well can final year primary school children develop relevant cybersecurity lessons for younger peers?

RQ2: What factors should be considered when using peer mentors in this context?

The remainder of this paper presents the results of an action research project within a Scottish primary school. This is followed by a discussion of the findings and the definition of a framework for implementing peer mentoring for cyber security in primary schools.

3 Methodology

We employed an action research methodology, which seeks to address practical concerns within problematic contexts while simultaneously advancing scientific knowledge within a rigorously defined ethical framework (Rapoport, 1970). We applied a cyclical action research process (Susman & Evered, 1978) but here primarily report on general findings and lessons learned. We present a descriptive analysis of survey responses and exercise outcomes, incorporating both quantitative and qualitative data, to enhance the understanding of children's performance and engagement. This analysis is augmented by observational data and selected quotes to provide additional insight.

The study was composed of four parts. The first developed a pilot framework based on findings from a review of literature. The second investigated the effectiveness of current cyber security resources readily available for primary schools in educating peer mentors in basic cyber security principles. The third investigated the pupil-led development of learning plans and resources for younger years. The fourth and final part investigated the delivery of these lessons. This section reports on each of these parts in turn.

3.1 Part 1: Pilot Framework

A pilot framework consisting of three stages was developed to support the delivery of cyber security education from older peers to primary school aged children. The pilot framework is presented in Fig. 1. To distinguish between the different pupils involved in the study, the Primary 7 pupils are referred to as mentors, and the younger pupils are referred to as mentees.

Before the mentors could deliver cyber security education, it was important that they too were well-educated in cyber security constructs (Tondeur et al., 2017). Research has shown that the cyber security knowledge of pupils in the upper stage of primary school is

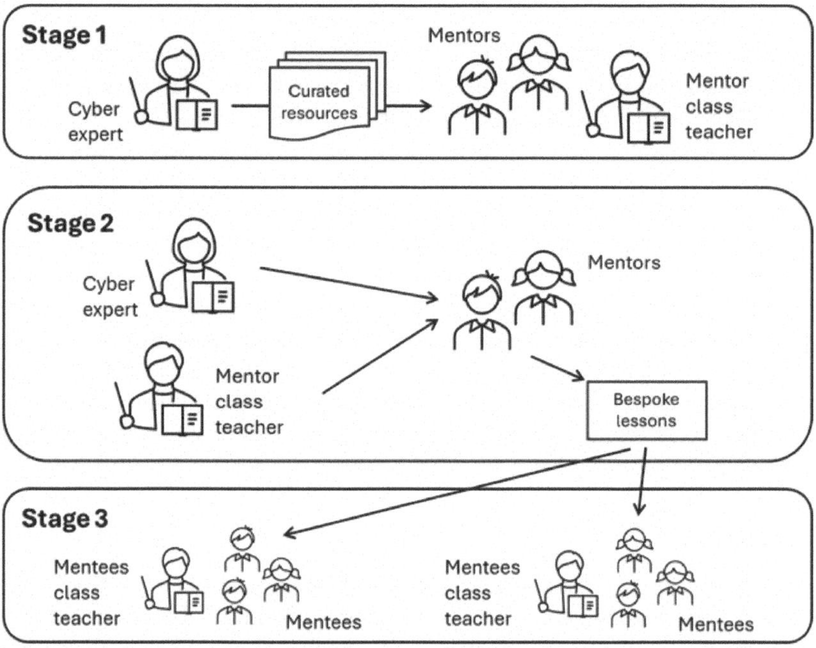

Fig. 1. Pilot Framework

mixed and often contains outdated principles (Prior & Renaud, 2022). Therefore, Stage 1 involves the mentors being educated by cyber security experts.

To deliver bespoke cyber security lessons, the pupils would need support in scaffolding activities. Stage 2 involves developing short lessons for younger years, which was supported by the class teacher and the cyber security experts. Following, Stage 3 consists of the delivery of the lessons to a younger group. In line with Karcharer's (2013) definition of cross-age peer mentoring, the mentors were at least two years older than the class being mentored.

3.2 Part 2: Training of Peer Mentors

The second part of this study concentrated on the training of peer mentors. A school with a need for cyber security education was identified as a partner in this study. It represented a diverse catchment area – while much of the catchment was an affluent area of city suburbs, in the top 20% of homes in Scotland (Scottish Government, 2020), over 10% of the school's children were from financially deprived backgrounds. Thus, the school presented a rich context with various child backgrounds.

The researchers' institutional Ethics Committee approved the study. Before the research started, both researchers had criminal background checks conducted. The head-teacher provided informed consent on behalf of the school and introduced the researchers to the class teacher.

All pupils in Primary 7 (generally aged 10–12 years) were invited to participate. Informed consent and information sheets were sent home with the pupils, informing parents that the researchers would conduct cyber security lessons within the school as part of the national curriculum over the academic year and that they could choose if they wished their child's data to be included in a research study. Parents were informed that the data would be pseudo-anonymized before analysis and stored securely. All data was stored securely separately from the consent forms from participants. Informed consent was obtained for 23 of the 43 children (53%) eligible to participate. Data was only retained at the point of collection for children for whom ethical consent had been obtained. All other responses were destroyed securely and are not reported on.

Over three sessions between November 2023 and January 2024, the researchers delivered the National Cyber Security Centre (NCSC) CyberSprinters[1] resources and lessons to the class. The materials were chosen due to the depth in which they covered key cyber security topics (Prior & Ophoff, 2024). The materials are designed to be used over the course of three lessons. Each lesson lasted approximately 90 min and took place in the mentors' classroom. The mentors' responses are considered against the intended learning outcomes of the lessons.

3.2.1 Passwords

The first lesson, 'Creating and managing passwords' has learning objectives centered on strategies for maintaining password privacy and explaining and creating strong passwords. At the start of the lesson, mentors were asked to answer five questions: *How do passwords work? What makes a strong password? Why do we need passwords for our online accounts? Why do email accounts need to be kept safe? Where can we get help around protecting our online accounts?*

Before the lesson, the mentors' average knowledge score was 52.6% (min = 20%, max = 80%). Post-lesson, the mentors scored an average of 73.6% (min = 40%, max = 100%).

3.2.2 Device and Information Security

The second lesson, 'Protecting your devices' has a learning objective focused on understanding why and how to keep information and devices secure. There were no formal pre-measures taken in this lesson, but following the topic on information sharing mentors were presented with four scenarios and were asked to identify if it was safe to share information in four situations: *Their full name and date of birth with a doctor. The school they attend on a gaming site. Their date of birth on in-game messaging. The country they live in on a shopping account.*

Mentors performed well in this activity with an average score of 88% (min = 25%, max = 100%).

3.2.3 Scams

The final lesson, 'Suspicious contact' has a learning object to identify and manage suspicious contact. These exercises focused on phishing, smishing, and vishing. While not in the lesson scope, there were several questions posed by the mentors as to how

[1] https://www.ncsc.gov.uk/collection/cybersprinters

these would relate to gaming scams. A pre-measure based on questions regarding an example suspicious message was taken. A sample phishing message was displayed on the screen, and mentors were asked four questions: *What type of suspicious message is this an example of? What are the signs in this message that make it suspicious? What other examples of suspicious messages are there? What can someone do, or where can they get help if they think they have received a suspicious message?*

In the pre-measure, mentors scored an average of 53.3% (min = 25%, max = 75%), and in the post-measure, mentors scored an average of 68.3% (min = 25%, max = 100%). Following these lessons, it was felt that mentors had an appropriate level of understanding of the basic cyber security concepts required for children of this age, according to the materials used.

3.3 Part 3: Creation of Lessons

Upon completing the three NCSC CyberSprinters lessons, the third stage of this study involved the mentors selecting a topic and target audience and developing an activity for this group.

3.3.1 Year-Topic Lesson Allocation

Mentors were given a form listing the different pupil year groups within the school and the cyber security concepts covered in CyberSprinters, and asked to select any that interested them. Mentors were instructed to consider from their own experience what topics would have been useful to them at different ages. This allowed us to gauge which year groups mentors wanted to work with, and which specific cyber security topics they felt to be important for those groups.

Based on this, groups of three to four mentors were formed. All mentors were given one of their selected topics and year groups to work with. As seen in Fig. 2, the most popular choice was in the youngest year, Primary 1 (4–5 years), and the topic of passwords. Device and information security was the least popular choice overall. For the older years the focus tended to move to scams.

3.3.2 Development of Thinking

Two sessions were held with the researchers observing and supporting mentor groups where necessary, as they worked on their lesson/activity development. Groups were given the following points to consider and help them focus on a coherent lesson:

1. *What class are you working with? What do you think they might already know about the topic?*
2. *What would you have liked to know at this age?*
3. *What are you going to teach them? Remember that it might be easier to focus on one part of the topic.*
4. *What activity will you do with the group? Think about things such as how much they can read and write on their own.*
5. *How will you explain the topic to them before you start?*
6. *What instructions will you give them about the activity?*

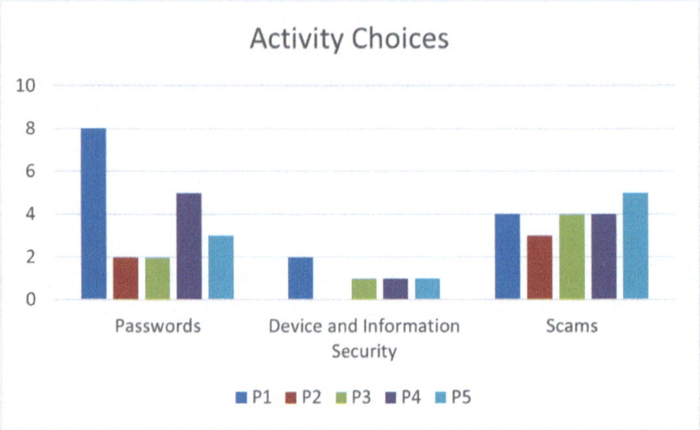

Fig. 2. Mentors' Preferred Choices

7. *How will you finish the lesson?*

To illustrate the children's experience of this process responses to some of these questions are included below, along with reflections based on the researchers' observational notes of working with the groups.

Groups working with the youngest years in the school felt that, in general, these classes knew little regarding passwords and emails. As the age of the years being worked with increased, the sense from the groups was that the general level of knowledge would be higher, even when considering the more advanced topics being discussed. For those working with year groups in the middle of the school (7–9 years), the groups felt they would know about the need to keep accounts and information secure from others, that passwords should be secret, and that they would have some awareness of the term 'online scam'. The groups working with the eldest pupils (9–11 years) felt that these mentees would be able to define the term 'phishing' and would also be aware of the dangers involved while being online.

In general, there was agreement about what the mentors would have wanted to know when they were younger. Those working with the youngest years felt that they would have liked to know the fundamentals of password security, a sentiment shared by all groups, for example:

> *"How to make a secure password and who to tell your password and how to keep your password secure."* (written response)

Within one group discussion arose around how they felt that children aged 5–6 years should be shielded from the internet:

> *"P1/P2 shouldn't even be on the internet because it messes with their mind."* (comment from mentor recorded in notes)

Most groups felt less strongly about this and that it was acceptable for this age group to be online. However, a point was raised about informing parents (or a trusted adult) about online activity:

"For them at their age they should be sharing everything with their parents." (comment from mentor recorded in notes)

It was observed that the focus moved from the fundamentals of security for younger years, to a focus on cyber safety for the middle years of the school. While one group still felt it was essential to learn about who to share a password with, the other groups raised the issues of stranger danger and "*reporting weird behavior*" (written response). This observation was also valid for those working with the oldest groups who, in addition to feeling the need for guidance on what to do if their information was stolen, would want to know "*how to tell the difference between hackers and normal people*" (written response).

Within their lesson plans most of the groups began by considering some form of active learning that could take place with the class. This ranged from a quiz to.

"Guess the password hangman" and

"Sellotape questions on a ball, then when they catch it, they read out a question, and someone answers." (written response)

Many groups also included a (PowerPoint) presentation to explain the topic or define some of the security terms for the class. Some groups put a great deal of thought as to how they would explain the concepts in a simplified way to a particular year group, for example:

"Describe a padlock and a key and say that computers are like that too and the key is called a password and that you need them so that like a padlock you can stop people stealing your stuff and computers are much the same." (written response)

Groups working on passwords were able to create concept definitions and explanations quicker than those who were conducting lessons on other security concepts, such as 'a scam in an email'. These groups needed more support with developing definitions which did not rely on prior knowledge from mentees.

The details in the instructions planned for the activities did not differ between the year groups being worked with. However, some groups specified only minimal detail on what instructions would be given, e.g. *"put 3 words together to make a strong password"* (written response). In contrast, other groups created a script for each section of the lesson:

"Intro: so <name> is going to think of a password and give you a couple of hints. Do you think you can guess it?" (written response)

Except for one, all groups described a plan to end with a quiz to ensure the lesson was effective. The majority included a recap to summarize the learning objectives. Only one group suggested offering the class the opportunity to ask their own questions about anything they were unsure about.

Following two sessions with the researchers on hand to offer guidance on the cyber security topic content, the mentors continued to develop the lessons over three weeks with the class teacher's support, including the suggestion that they all incorporate a video explaining the topic into their lesson.

3.4 Part 4: Delivery of Lessons

Once the lessons were finalized the class teacher arranged for the mentors to carry out the lessons with the younger years. This took place over two days to fit into the school schedule. During the lesson the mentees being taught had their teacher present, and the researchers assumed a non-participatory observational role. Apart from introducing themselves to the class teacher, and explaining what would happen, there was no further researcher involvement with the delivery of lessons. The researcher sat behind the class and did not engage with mentors or mentees.

The class teacher was invited to participate in the research and given an informed consent sheet. No data was collected from any mentees in the classes being taught. At the end of the activity, the class teacher was invited to complete a short questionnaire on their experience: *Do you think your class learnt anything new during the activity? Do you think your class enjoyed or got excited about the topic? Do you feel you could carry out this topic yourself? Do you think your class should learn more about this at this age? Do you think the use of P7 pupils in delivering the activity was effective?*

Three teachers completed this and their responses are used alongside researcher observations to report on the effectiveness of using peer mentors.

3.4.1 Learning

The three teachers who responded to the questionnaire reported feeling that their class learned something new. In sessions with younger classes covering passwords, it was observed that the knowledge the mentees possessed was outdated, such as complex LUDS (Lowercase Uppercase Digits Special Characters) in connection with passwords. The peer mentors confidently encouraged the mentees to consider whether this would be memorable.

3.4.2 Enjoyment

In all classes there was great enthusiasm for the peer mentors to participate in the lesson. Despite some classes being very excitable, the peer mentors were able to ensure that everyone took part in the lesson. In most cases, the peer mentors were confident in tackling minor behavioral challenges using school behavior approaches such as repeated clapping of a pattern to gain attention. It was observed that learning seemed most effective in classrooms where the teacher also took an active role in managing mentees' attention.

While mentees paid attention to the videos shown during the lesson, the most enthusiasm was seen when participating in the active learning activities. Teachers said the class enjoyed the activities and learning about the topic.

3.4.3 Teacher-Led

Two teachers responded to the question whether they could carry out this topic themselves. Both felt they would be able to carry out the activity, and one offered suggestions as to how it could be augmented. In the sessions, there were occasions when the teachers offered additional input to ensure that the class understood the lessons. Two teachers commented that they did not know some of the answers and were unaware of current guidance on topics such as passwords.

3.4.4 Peer Mentors

When working with the youngest years, it was noticed that the mentees in those classes were keen to learn from the peer mentors and viewed them as educators, raising their hands and adhering to instructions. When mentees in the class asked questions, the peer mentors were confident in responding and could provide the correct guidance. In the youngest classes, this was often supported by the teacher, who asked additional questions to ensure the mentees had understood the messages and provided additional examples were necessary.

During lessons the mentees who were closer in age to the mentors asked more challenging questions. This could have been due to their greater understanding and interest in the topic or could have been an attempt to display greater knowledge than the mentors. In one older class, it was noticed that the peer mentors were being challenged on their responses, and in this case, the teacher stepped in to confirm the peer mentors were giving the correct information.

3.4.5 Classroom Aspects

The enthusiasm of the class teacher appeared to significantly impact the overall experience of the lesson being delivered. Classes in which the teacher showed interest, asked questions, and removed the burden of behavior management from the mentors appeared to be more successful in completing the lesson overall.

The general mood in the classroom also appeared to impact the overall experience. On one of the days, it was noted that it had been a 'wet lunch' – a lunch break in which all pupils were kept inside the classrooms, with the older pupils in the school monitoring the behavior and activities of the younger ones. On this day, the mentees had a lot of pent-up energy, and maintaining their focus on the lesson was more challenging than on the other day in which a normal lunch break had taken place.

4 Discussion

It has long been recognized that children need to receive cyber security education, and there are many resources available for those tasked with delivering this education (Lamond et al., 2022). There has been less focus as to who is best placed to deliver these lessons – indeed one criticism of academic interventions is that they frequently exist outside of a typical school setting (Ondrušková & Pospíšil, 2023). This study aimed to address this criticism by conducting research within a typical classroom using existing cyber security educational resources.

4.1 Returning to the Research Questions

This study posed two research questions, which are considered here, considering the results of the study.

The findings were generally positive regarding the question: *How well can final year primary school children develop relevant cybersecurity lessons for younger peers?* (RQ1) In developing the activities the mentors considered the age and knowledge of the target class. The mentors often chose active lessons with practical activities, reflecting findings from Nicholson (2021) that young people wanted more hands-on activities in cyber security. The activities chosen were relevant and, in some cases, highly original with a strong message. It was positive to see how mentors were able to select aspects of a topic and focus on this for their target class.

We identified several issues regarding the question: *What factors should be considered when using peer mentors in this context?* (RQ2) Empowering the mentors to share their own experiences or explain their own practice, which the mentees could mirror, may have benefits (Kohut, 2009) but would need to be carefully managed. In the CyberSprinters lessons it was observed that mentors often tried to draw connections between the topic examples, such as phishing emails, and their own experiences of scams through in-game messaging. Clear guidance for those supporting the mentors in managing this would be of benefit. In addition, the role of the mentees' class teacher cannot be overlooked. It was observed that while the mentors were confident in delivering the lessons and in managing low-level disruption in the classroom, the more successful lessons occurred when the class teacher was taking an active role in managing class behavior and in asking questions or clarifying points to ensure their class was listening and understanding the points made.

4.2 Limitations

A limitation of this study is that we were unable to request informed consent from all children in the school due to admin constraints. This would have enabled assessing the younger pupils' pre- and post-knowledge as well as allowing us to ask them directly about their experience. In lieu of this we asked the expert in the room – the teacher. Any future study examining the proposed framework would benefit from the inclusion of this 'recipient' group of participants.

5 Proposed Cyber Peers Framework

Based on the findings of this research a refined framework, entitled Cyber Peers, has been developed. The framework consists of four stages and is illustrated in Fig. 3.

Stage 1 – Education of the Educator

While the inclusion of cyber security experts in the pilot undoubtedly had benefits, it is unrealistic for this to occur in every class. Instead, it is proposed that one cyber security expert deliver a short lesson on the process and examples of how existing resources can be utilized for digital leads within schools. Digital leads are already present within most

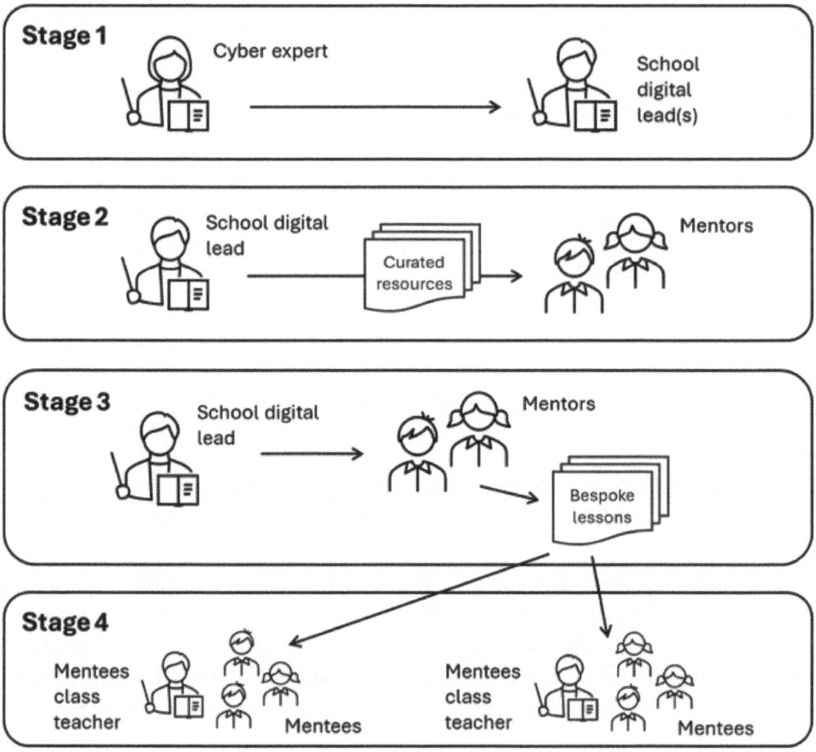

Fig. 3. Cyber Peers Framework

schools and are responsible for supporting other teachers within the school in their own digital practice and in the digital education of the pupils (Glow, 2023). The digital lead would be well positioned as a future cyber topic expert within the school.

Stage 2 – Education of the Mentors

Once a digital lead has been trained in the implementation of Cyber Peers, they would then take responsibility for overseeing its rollout within their school. The digital lead would work with a small group of senior pupils who have expressed an interest in the project. This is like existing projects within schools in which older pupils take leadership roles on topics such as reading, the environment (Keep Scotland Beautiful, 2023), and active travel (Mode Shift Stars, 2019). Within existing schemes, these training sessions often occur during lunch breaks. Given the smaller number of mentors it would be possible to deliver this learning in shorter periods of time. Schools already have access to many different cyber security resources, and these sessions would look at using these to educate the mentors and consider how they might be applicable for the mentees.

Stage 3 – Preparing to Educate Mentees

With support from the digital lead, the mentors should develop relevant activities for the mentees. This may involve utilizing existing resources but could also, as was seen in this research, involve developing new activities or tailoring the learning to the issues facing children at this time, or in a specific context.

Stage 4 – Education of the Mentees

The final stage of the framework is the education of the mentees. In a similar fashion to this research, the mentors would deliver activities and lessons to the younger children. In addition, clear guidance would be given to class teachers explaining how they can support the mentors in delivering the lessons. The class teacher should become an active participant in the lesson delivery.

6 Conclusion

This study investigated the feasibility of peer mentors being used with young children in cyber security education. The results of a pilot study indicate that with the correct guidance and support from both those educating the mentors and those supporting them in educating the mentees, there can be benefits for both parties. We propose a framework for delivering this scheme more widely. Future work in the area should consider how best to carry out development training for the digital leaders in schools. It would also be worthwhile to examine the extent to which the younger children learnt new information during the sessions. It is also not yet known what impact these sessions have had on the actual actions of the mentors and mentees – while they may have acquired new knowledge more work is needed to determine to what extent this is implemented in their cyber practice.

Acknowledgments. We thank the school for allowing us access to conduct the research.

Ethics Approval. This study was approved by the Abertay University Ethics Committee (EMS7231).

References

Childers, G., Linsky, C.L., Payne, B., Byers, J., Baker, D.: K-12 educators' self-confidence in designing and implementing cybersecurity lessons. Comput. Educ. Open **4**, 100119 (2023). https://doi.org/10.1016/j.caeo.2022.100119

Craigen, D., Diakun-Thibault, N., Purse, R.: Defining cyber-security. Technology Innovation Management Review. **4**(10), 13–21 (2014). https://doi.org/10.22215/timreview/835

Education Scotland. Benchmarks Technologies (2017)

Glow. This is… Digital Leaders: skills to help you lead change in your school. Digi Learn Scotland (2023). Retrieved 11 June 2024, from https://blogs.glowscotland.org.uk/glowblogs/digilearn/2023/03/27/this-is-digital-leaders/

Grey, A.: Cybersafety in early childhood education. Australas. J. Early Childhood **36**(2), 77–81 (2011). https://doi.org/10.1177/183693911103600210

Karcher, M.: Cross-age peer mentoring, pp. 233–258 (2013). https://doi.org/10.4135/9781412996907.n16

Kaspersky. Overconfident and over exposed: Are Children Safe Online? Children are going online like never before. But just how safe are they? And who is helping protect them from cyber threats? (2023)

Keep Scotland Beautiful. (2023, March). Eco-Schools. Keep Scotland Beautiful. Retrieved 11 June 2024, from https://www.keepscotlandbeautiful.org/eco-schools/

Kohut, H.: The restoration of the self. University of Chicago Press (2009)

Kumar, P.C., Chetty, M., Clegg, T.L., Vitak, J.: Privacy and security considerations for digital technology use in elementary schools. In: Paper presented at the Proceedings of the 2019 CHI Conference on Human Factors in Computing Systems, Glasgow, Scotland UK, pp. 1–13 (2019). https://doi.org/10.1145/3290605.3300537

Lamond, M.L., Renaud, K.V., Wood, L.A., Prior, S.: SOK: Young Children's Cybersecurity Knowledge, Skills & Practice: A Systematic Literature Review. ACM (2022). https://doi.org/10.1145/3549015.3554207

Lipka, R.P., Brinthaupt, T.M.: The role of self in teacher development. Suny Press (1999)

Mode Shift Stars. (2019, Active travel resources for teachers in Scotland. Mode Shift Stars. Retrieved 11 June 2024, from https://modeshiftstars.org/ata/

Muir, K., Joinson, A.: An exploratory study into the negotiation of cyber-security within the family home. Front. Psychol. **11**, 508279 (2020)

Nicholson, J., Terry, J., Beckett, H., Kumar, P.: Understanding young people's experiences of cybersecurity. In: Paper presented at the Proceedings of the 2021 European Symposium on Usable Security, pp. 200–210 (2021)

Nicholson, L.J., Rodriguez-Cuadrado, S., Woolhouse, C.: Reframing peer mentoring as a route for developing an educational community of practice. Mentoring Tutoring Partnership Learn. **26**(4), 420–440 (2018). https://doi.org/10.1080/13611267.2018.1530163

Ofcom. Children's Online User Ages Quantitative Research Study. ().Yonder Consulating (2022)

Ondrušková, D., Pospíšil, R.: The good practices for implementation of cyber security education for school children. Contemp. Educ. Technol. **15**(3), ep435 (2023)

Prior, S., Ophoff, J.: Lessons learnt from using educational phishing materials with UK primary school children. In: Drevin, L., Leung, W.S., and von Solms, S. (eds.) Information Security Education - Challenges in the Digital Age, pp. 36–49. Springer Nature Switzerland, Cham (2024). https://doi.org/10.1007/978-3-031-62918-1_3

Prior, S., Renaud, K.: The impact of financial deprivation on children's cybersecurity knowledge & abilities. Educ. Inf. Technol. **27**(8), 10563–10583 (2022)

Prior, S., Renaud, K.: Are UK parents empowered to act on their cybersecurity education responsibilities? In: Paper presented at the HCI International 2024 (2024)

Rahman-Jones, I., Vallance, C.: Online Safety Bill: divisive internet rules become law. BBC News (2023). Retrieved June 5 2024, from https://www.bbc.co.uk/news/technology-67221691

Rapoport, R.N.: Three dilemmas in action research: with special reference to the tavistock experience. Hum. Relat. **23**(6), 499–513 (1970). https://doi.org/10.1177/001872677002300601

Renaud, K., Prior, S.: The "three M's" counter-measures to children's risky online behaviors: mentor, mitigate and monitor. Inf. Comput. Secu. **29**(3), 526–557 (2021)

Scottish Government. Scottish Index of Multiple Deprivation 2020 (2020). http://www.gov.scot/collections/scottish-index-of-multiple-deprivation-2020/

Susman, G.I., Evered, R.D.: An assessment of the scientific merits of action research. Adm. Sci. Q. **23**(4), 582–603 (1978). https://doi.org/10.2307/2392581

Thurston, A., Roseth, C., Chiang, T., Burns, V., Topping, K.J.: The influence of social relationships on outcomes in mathematics when using peer tutoring in elementary school. Int. J. Educ. Res. Open **1**, 100004 (2020)

Tondeur, J., van Braak, J., Ertmer, P.A., Ottenbreit-Leftwich, A.: Understanding the relationship between teachers' pedagogical beliefs and technology use in education: a systematic review of qualitative evidence. Education Tech. Research Dev. **65**(3), 555–575 (2017). https://doi.org/10.1007/s11423-016-9481-2

Turner, S., Pattnaik, N., Nurse, J.R., Li, S.: You just assume it is in there, i guess: understanding UK families' application and knowledge of smart home cyber security. Proc. ACM Hum. Comput. Interact. **6**(CSCW2), 1–34 (2022)

UK Council for Internet Safety. Education for a Connected World -2020 edition (2020)

UK Government. Cyber resilience (2024). https://www.gov.uk/government/collections/cyber-resilience

Vygotsky, L.S., Cole, M.: Mind in society: Development of higher psychological processes. Harvard university press (1978)

Witt, M.: An evaluation into the efficacy of the FunKey Mathematics Peer Mentoring Programme (2020). https://www.funkeymaths.com/assets/uploads/2020/08/FunKey-Evaluation-Report.pdf

Experiences of a Pan-Wales Initiative of Cybersecurity Education in Schools

Megan Venn-Wycherley[1](✉), Joanne Ralph[2], and Faron Moller[1]

[1] Department of Computer Science, Swansea University, Swansea, UK
{megan.venn-wycherley,f.g.moller}@swansea.ac.uk

[2] CyberFirst, Swansea, UK
joanne.ralph@technocamps.com

Abstract. In the following chapter, we examine the experience of implementing the CyberFirst programme in Wales, drawing from experiences of delivery officers, ambassadors, teachers, industry partners, and young people. Through four case studies, we explore the challenges and opportunities in engaging young people, supporting teachers and collaborating with industry to develop an effective cybersecurity ecosystem. The chapter highlights the importance of tailoring content to young people's interests and experiences, providing accessible professional development for teachers, and bridging the gap between industry expertise and educational needs. Key insights include the need for pedagogical innovation in cybersecurity education, the critical role of continuous professional development and the potential of industry collaborations to provide authentic learning experiences. While focused on experiences in Wales, the chapter serves as a call to action for the broader cybersecurity education community, encouraging reflection on current practices and advocating for innovative approaches to address the growing demand for cybersecurity skills and to inspire a diverse future workforce.

1 Introduction

In today's increasingly digital world, cybersecurity has become a concern that transcends both geographical and geopolitical boundaries. From ransomware attacks on critical infrastructure, to massive data breaches compromising personal information, the impact of increasingly sophisticated cyberthreats is far-reaching, affecting individuals, organisations and nations alike. The demand for cybersecurity professionals who are able to address such vulnerabilities continues to outpace our potential to supply and train such a workforce (Furnell, 2021). In the UK, this critical workforce gap is particularly pronounced, with 50% of businesses facing a basic cybersecurity skills shortage and 33% reporting a deficit of advanced skills (Coutinho et al., 2023). With such stark statistics, focus has shifted towards the development of cybersecurity skills across all levels of the educational system, in order to encourage the development of a strong cybersecurity ecosystem.

A review of current of cybersecurity education in UK schools presents a loose curriculum with a tendency to focus more on e-safety, rather than well-rounded cybersecurity skills and concepts (Department for Education, 2013; Ofsted, 2022). This is attributed

P. Legg et al. (Eds.): CSE 2024, LNNS 1213, pp. 153–169, 2024.
https://doi.org/10.1007/978-3-031-77524-6_9

to a lack of specialist content knowledge, resource and infrastructure available in school environments (Nicholson et al., 2021; Saglam et al., 2023). In 2016, the CyberFirst programme was introduced to support awareness and skills development for young people interested in cybersecurity, sparking interest in relevant skills development, and signposting career pathways. Through government-supported extracurricular provision for young people via bursaries, girls-only competitions, and educational outreach, the CyberFirst programme aims to provide a cohesive programme of cybersecurity education and enrichment across the UK, addressing skill development and the future talent pipeline.

At a time when the UK Government is seeking consultation on the future of CyberFirst (Department for Science, Innovation & Technology, 2024), exploring input on the delivery and scale of the programme, as well as the role of government support, the following chapter examines the experiences of those facilitating the CyberFirst programme in Wales. Through a series of four short case studies, we provide an overview of the lessons learned and challenges experienced on the ground, painting a picture of the programme's impact on young people, the importance of exploring novel pedagogical approaches, and the role collaborations can play in improving cybersecurity education for young people. Together, these reported experiences serve as a call to action for the cybersecurity education community in the UK, and encourage a wider reflection on what can, and *should*, be done to achieve a more resilient digital society.

2 Background

In the following sections, we provide an overview of the motivations and current approaches to cybersecurity education in schools, before turning towards the role of the CyberFirst programme in developing and delivering cybersecurity education in young people. We explore the perceptions that young people hold about cybersecurity, how this influences the way they engage with educational initiatives, and the challenges of delivering effective cybersecurity education to young people. By understanding these perceptual and educational challenges, cybersecurity educators may better tailor their initiatives to capture the imagination and aspirations of young people, supporting the development of a more diverse, robust cybersecurity workforce and society.

2.1 Young People's Perceptions of Cybersecurity

Understanding, and shaping, young people's perceptions of cybersecurity is crucial when developing educational initiatives, which can significantly impact the way they approach cybersecurity as both a skill and a career path. A primary obstacle to engaging young people with cybersecurity is overcoming their misconceptions and internalised stereotypes, often influenced by popular media, which can portray cybersecurity professionals and their skills in a narrow light – as lone, expert, malicious actors (Fisk et al., 2023; Shires, 2020). These popularised portrayals can deter young people, who don't identify with these stereotypes, from exploring the relevance of cybersecurity skills and careers to their own identities and aspirations (Saglam et al., 2023). The impact of these misconceptions is particularly pronounced in areas of socio-economic disadvantage, where under-resourced young people may have had limited exposure to positive, real role models in the cybersecurity field and lack opportunities to learn about diverse career pathways (Mountrouidou et al., 2019).

Additionally, industry collaborations can allow students to see beyond the stereotypes and understand the dynamic nature of cybersecurity work – from threat analysis to network security, from policy development to digital forensics. However, it is important to note that establishing and maintaining such collaborations can be complex, requiring sustained effort. Educational institutions may face challenges in identifying willing industry partners with the skills to ensure that the experiences provided are both meaningful and relevant. Similarly, industry partners may face challenges in allocating resource and time to support these initiatives alongside their primary operations (UK Government, 2021).

Another significant challenge is the age at which young people typically engage with cybersecurity concepts (Saglam et al., 2023). Often, a formal introduction to cybersecurity concepts beyond e-safety occurs too late in a student's educational journey. By the time cybersecurity is presented as a potential career path or area of study, many young people have already formed their career aspirations and committed to other educational paths, having not selected pre-requisite courses required to pursue cybersecurity. Additionally, those who might excel in cybersecurity may have already internalised misconceptions about cybersecurity, viewing it as overly technical, male-dominated or inaccessible, particularly impacting those who are already underrepresented within the field (UK Government, 2021; Saglam et al., 2023).

2.2 Challenges of Cybersecurity Education for Young People

Building upon this understanding, it is crucial to consider the role of interactive and engaging educational experiences in shaping young people's perceptions of cybersecurity, particularly at early stages of their educational journeys. Traditional classroom-based learning, while foundational, often falls short in capturing the dynamic, rapidly evolving nature of cybersecurity – usually visited as a one-off lesson at specific points in the year (Nicholson et al., 2021). Attributed to limited resources, poor opportunities for teacher upskilling and lack of clear curriculum, schools often not in the position to deliver high quality, industry-relevant cybersecurity education (Saglam et al., 2023).

Cybersecurity is lightly touched upon in schools, starting with e-safety at a primary school level and moving towards a more complex understanding of threats and attacks when young people enter secondary school (Department for Education, 2013). Many of the nuanced cybersecurity skills and understanding is at the elective stages of formal education, which have limited representation of young women and socio-economically disadvantaged young people, as they are less likely to continue their computing studies (Kemp et al., 2024). This has a direct impact on the diversity of young people who are able to develop the skills and understanding to transition to further study and employment in the cybersecurity field.

Establishing resources that strike a balance between theoretical concepts and practical, ands-on education is crucial to avoid outdated or inadequate learning outcomes. However, delivering cybersecurity education in school environments is particularly challenging. For example, keeping up to date with the rapidly evolving landscape of cyber threats and technologies is challenging within school settings (Chase & Uppuluri, 2022), where digital infrastructure, software and training materials, as well as access to ongoing training for continuous professional development (CPD) can be limited (Fisk et al., 2023; Saglam et al., 2023).

Drawing insights from the field of computing education as a proxy for limited cybersecurity education literature focused on schools, we can understand that teachers are concerned that computing education within schools tends to be *taught to the exam*, meaning that the content of exam board specifications drives what is taught within school. With limited timetabled hours and the pressure to achieve results, teachers report that any additional content is often avoided in order to focus on assessed knowledge (The Royal Society, 2017). Moreover, while resources for computing education may not be difficult to find, educators often struggle to discern which materials are of good quality and can be trusted (The Royal Society, 2017). Understanding these challenges through the lens of cybersecurity, with the lack of a centralised, examined curriculum and a rapidly evolving landscape where resources quickly become outdated, presents another layer of complexity to the realities of delivering cybersecurity education effectively within the classroom (Nicholson et al., 2021; Saglam et al., 2023).

A further challenge is that of educator training, especially when considering the nuances of effective delivery of cybersecurity education to young people. Across the UK there is a notable shortage of qualified computing educators (Ofsted, 2022) and even fewer with relevant cybersecurity qualifications, particularly at the school level, when positions in industry often provide more lucrative opportunities with better working conditions. As a result, schools struggle to attract and retain educators with the necessary expertise to effectively teach cybersecurity concepts.

The problem is further compounded by the limitations of current CPD approaches in computing-relevant disciplines. The predominant model for teacher CPD is often criticised for its deficit approach, rather than a holistic, contextually aware approach towards teacher development (The Royal Society, 2017). Teachers who attend CPD sessions report that the content presented can be difficult to implement in their own, local context (Neutens & Wyffels, 2018). This creates a cyclical problem: a shortage of qualified educators leads to inadequate, patchy cybersecurity education, which in turn fails to inspire the next generation of potential professionals and educators.

2.3 The CyberFirst Programme

In 2015, the UK Government began to lay the groundwork for a comprehensive approach to engaging young people with cybersecurity education and careers (Cabinet Office et al., 2016). Initial educational efforts involved targeted summer programmes designed to support 14–18-year-olds in developing relevant cybersecurity skills and aspirations, with further focus on developing teacher skills and embedding cybersecurity into wider educational engagements (UK Government, 2021).

It was against this backdrop that the CyberFirst programme emerged in 2016 as a government sponsored initiative developed in collaboration with QA, designed to offer a range of opportunities for young people interested in cybersecurity (UK Government, 2021). The Discovery Camps programme was its first iteration, as an annual event for young people to explore cybersecurity opportunities and skill development. While this programme was successful in engaging young people, it showed limited impact on influencing long-term career aspirations (UK Government, 2021), attributed to the fact that many of the young people who chose to attend did so as they were already interested in cybersecurity opportunities for the future.

Building upon these insights, CyberFirst adopted a longitudinal, age-based approach to promote progression through the talent pipeline, that allows for flexible development of content alongside the changes in aspirations for young people on their educational journey. The programme begins with "Trailblazers" a half-day event for 11–12-year-olds, that provides hands-on learning through activities such as *Digital Detective*, where young people use digital forensic techniques to identify the originator of a contagious disease and use open-source intelligence to discover more information about this person. This is followed by "Adventurers", another half-day event for older students, focusing on career pathways just as they are making decisions about their GCSE choices. These earlier interventions aim to challenge stereotypes and demonstrate applications of cybersecurity that can be difficult to authentically present in schools.

As students progress, they participate in more intensive CyberFirst courses – both in duration and content. "Defenders" is a five-day course for 14–15-year-olds, offering practical skills in threat identification, personal cyber hygiene and the deployment of front-line cyber defences. "CyberFirst Futures" aimed at 15–16-year-olds, explores the human behaviour elements behind cybersecurity, such as motivations for cyberattacks. For 16–17-year-olds, "CyberFirst Advanced" provides a five-day deep dive into advanced skills such as digital forensics, encryption technologies, open-source intelligence techniques and penetration testing. Despite the limited representation of cybersecurity in the National Curriculum, all these activities are designed to complement the knowledge and skills required of the computing programme of study.

The CyberFirst programme culminates with Higher Education support, offering financial bursaries for university students, alongside summer placements, and access to professional events and communities. Additionally, the degree apprenticeship route provides a three-year apprenticeship opportunity to earn while they learn, in which young people gain exposure to cutting-edge technologies along with practical insights into their application, while they study the conceptual and theoretical elements of cybersecurity. This is intended to create a clear pathway from formal education to a career, with multiple touchpoints with industry.

In addition to this core offering, CyberFirst has developed specialised initiatives, like the CyberFirst Year 8 Girls competitions, designed to address the lack of female representation in cybersecurity, through a mix of challenges that cover topics such as cryptography, AI and networking. CyberFirst also offers Development Days, one-day courses starting from Year 9, designed to increase understanding of the diverse, wide-ranging specialisms within cybersecurity.

The impact of the scheme has been significant, with evaluations noting the positive changes in career perceptions among participants, regardless of their ethnicity, gender or background (UK Government, 2021), leading to a marked increase in the number of young people expressing interest in pursuing cybersecurity careers. However, the development and implementation has not been without challenge, particularly in reaching underrepresented young people and making the educational impact accessible to those who might not have the resources or access to attend.

Building on the lessons learned from the broader CyberFirst initiatives, the CyberFirst Schools scheme was developed to create a sustainable and embedded approach to cybersecurity education with schools (Cabinet Office et al., 2016). This scheme aims to recognise and support the schools that demonstrate a strong commitment to excellence in cybersecurity education, who are then able to act as a hub to support student engagement. Schools can apply for CyberFirst School status, with recognition not only validating the school's efforts in promoting cybersecurity but also providing additional resources and opportunities for both students and teachers. The scheme supports achievement of career-based OFSTED requirements, provides access to the CyberFirst Industry companies network, and ensures support from regional CyberFirst ambassadors.

The CyberFirst ambassador programme can play an important role in providing sustainability and scalability to the delivery of cybersecurity education. By leveraging industry-based ambassadors the programme creates a bridge between schools and industry, offering young people authentic insights into potential careers and pathways. These ambassadors provide support at a regional level, through inspirational engagements sharing their experiences and career journeys, e-mentoring to provide ongoing support and guidance to young people, and facilitating work experience opportunities that give young people hands-on exposure to the industry in action. Additionally, ambassadors contribute to the expansion of cyber activities by supporting coding clubs and extracurricular initiatives, helping to embed cybersecurity concepts into educational offerings. This approach not only enhances the reach and impact of cybersecurity education but can also ensure that content remains current and relevant to industry needs (UK Government, 2021).

2.4 CyberFirst Schools in Wales

In 2023, the Welsh Government outlined a *Cyber action plan for Wales*, which proposed four priority areas for cybersecurity: growing the cyber ecosystem, building a pipeline of cyber talent, strengthening cyber resilience and protecting public services (Welsh Government, 2023). Noting that the Welsh context provides its own challenges and opportunities, the policy outlined the importance of developing digital and cyber skills to support all to build a more equal, more resilient Wales. This is exemplified through Welsh Government support of early pilot projects alongside CyberFirst, resulting in the wider CyberFirst programme in Wales. At the heart of the current CyberFirst programme in Wales, is a partnership between three prominent Higher Education Institutions: The University of South Wales, Swansea University, and Bangor University. With the University of South Wales spearheading the programme within the region, outreach partners Technocamps and the National Digital Exploitation Centre were introduced to provide wide geographical reach and diverse expertise. The experiences within this chapter are drawn primarily from delivery officers within the Technocamps organisation, a pan-Wales STEM outreach organisation led by Swansea University. With a track record dating back to 2003, Technocamps has been instrumental in providing computing skills and experiences to thousands of young people in Wales every year. Their involvement in CyberFirst has been crucial in leveraging existing relationships and resources to deliver trusted content across Wales. They have provided a vital role in supporting schools to achieve CyberFirst accreditation, by helping them meet the minimum hours of enrichment activities through engagement with their computing outreach offering. The impact of the CyberFirst programme in Wales is evident, with over 500 students completing CyberFirst Adventurers since the launch of CyberFirst Wales. Twenty schools and colleges across Wales have achieved CyberFirst School Awards, marking a significant milestone in the recognition of excellence in cybersecurity education in Wales. The programme has also shown a particular focus on encouraging girls to enter the field, with more than 1,100 Year 8 girls participating in the CyberFirst Girls competitions in 2023, and over 150 Year 10 girls attending events that promoted computing A-Levels. Furthermore, the Welsh approach to CyberFirst, taking advantage of the regional approach to outreach delivery and engagement, has placed key emphasis on industry partnerships and regional relevance. One of the standout features of this approach is the "CyberFirst Wales Industry Partner" scheme. Schools and colleges that successfully achieve the CyberFirst award are paired with a local cyber business, establishing an annual engagement framework of exclusive events and activities in pursuit of their computing education strategies. The impact of these partnerships is demonstrable, with 13 businesses and 37 professionals contributing over 1,000 h of volunteering time to support the programme. This commitment underscores the programme's relevance to real-world cybersecurity needs, while also providing young people with valuable, authentic insights into potential career pathways. Over the past year, over 300 young people aged 16–18 have attended events specifically designed to help them identify routes into the cybersecurity sector, bridging the critical gap between education and employment. This presents a regional model that is deeply interested in creating integrated, personalised and locally relevant cybersecurity education experiences for young people.

3 Experiences

The following section explores the first-hand experiences of the individuals involved in the delivery of the CyberFirst programme in Wales, drawn from delivery officers, ambassadors, teachers, industry partners, and young people. Through the lens of four case studies, we demonstrate the impact of the initiative on young people's engagement with cybersecurity, the pedagogical development of activities, approaches to teacher support and the role collaboration can play in supporting such initiatives.

3.1 Case Study 1: Reframing Resources

The implementation of the new Curriculum for Wales presented an opportunity for the CyberFirst Wales team to restructure their educational delivery. The team recognised that for their content to align with cross-curricular learning experiences required by the new curriculum, they would need to transform the half-day Trailblazer programme into three distinct one-hour workshops. Drawing upon the pedagogical insights of Technocamps, delivery officers used their expertise to redesign QA resources to be more engaging to young people. The following case study examines the development and impact of these workshops: Lemming Pox, Fake News and Cryptography.

The Lemming Pox workshop, adapted from QA's Patient Zero module, introduces students to the concept of a viral social media phenomenon – in with a digital virus is infecting social media accounts and causing people to lose followers. In the original version of the content, students explore medical records to identify Patient Zero, whereas Lemming Pox is couched in the framing of social media – engaging young people by relating to their daily experiences and familiar terminology.

The workshop emphasises critical thinking and data analysis skills, where young people are challenged to compare facts and look beyond obvious conclusions. This is reflected in feedback from one young person, who noted: "*I enjoyed it because it challenged me and taught me to compare facts as the most obvious in not always the correct one.*" The use of spreadsheet work adds a practical, technical component that young people found engaging, with another participant noting, "*I really enjoyed the spreadsheet part of it, which was challenging but fun.*"

Recognising the prevalence of misinformation in today's digital landscape, with young people being some of the most exposed to AI generated content, the CyberFirst team developed a workshop exploring Fake News. The session begins with a topical discussion about deep fakes, using FKA Twigs' self-created deep fake to spark debates about the pros and cons of such technology, at the suggestion of young people from a previous Trailblazers workshop.

The workshop then provides hands-on experience with HTML manipulation, demonstrating how easily news headlines can be altered. This activity is carefully structured to ensure young people understand the ethical implications of these actions and culminates in young people creating their own fake news articles using Canva and AI-generated images, bridging technology with creativity in the creation of headlines such as "*Undiscovered species of robots found on Mars*" and "*Prehistoric Alexa reveals cavemen are smarter than we thought.*" Feedback from young people highlights the workshop's effectiveness in raising awareness about digital manipulation: "*It made me more aware of how easy and scary technology can be,*" with other students noting their appreciation of the creative aspects of generative technologies.

The Cryptography workshop was designed with language departments in mind, showcasing the interdisciplinary nature of cybersecurity. These sessions incorporated a range of problem-solving activities, from identifying lost European languages to using Pig Pen and Caesar ciphers. The introduction of the Cipher Chef tool provides young people with practical experiences in cryptographic techniques.

The workshop also introduces open-source Intelligence techniques, challenging young people to predict the next location of the fictional Lemming Pox outbreak, helping to conceptually tie the three workshops together. This combination of historical ciphers and modern intelligence gathering techniques offers a broad perspective of cybersecurity. Young people particularly enjoyed the collaborative and competitive aspects of the workshop, with one participant remarking, "*I really enjoyed working as a team and completing the challenges. It's really satisfying figuring out a puzzle.*" Another young person noted how they appreciated the fast-paced nature of the activities: "*I enjoyed this most because it was competitive and fast-paced!*"

3.2 Case Study 2: Teacher Conference

The annual Technocamps Teacher Training Conference serves as a crucial platform for digital educators across Wales to share best practices, stay informed about curriculum changes, and discover relevant opportunities for support. In 2023, the conference brought together over 30 educators from primary and secondary schools, with a focus on enhancing cybersecurity education in Welsh Schools. CyberFirst delivered the first part of the day, presenting the CyberFirst Schools programme, which outlined the various opportunities available to young people and schools in the field of cybersecurity. In the afternoon, the Technocamps delivery team conducted hands-on cybersecurity workshops, demonstrating how cybersecurity and digital forensics could be effectively taught in schools, running through the open educational resources with teachers in the position of their learners.

This approach to CPD was noted to be very well received by teachers, with one facilitator remarking that "*We go through (the resources) as we would teach them in the lesson. We feel that works best with teachers, so they can see how to present themselves to their classes.*" Through this practical, immersive method, teachers are able to experience content from a student's perspective and reflect on their own approach to delivering the resources in their own context, addressing some of the challenges noted in current CPD opportunities for teachers wanting to upskill in cyber-related fields. Teacher feedback from the event demonstrated how this direct access to the network itself was important, allowing teachers the opportunity to *"[speak] with others about the purpose of CyberFirst,"* as well as a *"greater understanding of the pathways into Cyber, the barriers and how we can try to overcome them."* When asked what to improve about the event, teachers asked for further opportunities to train or upskill in order to attract more teachers, as well as allow for further opportunities for networking.

The conference not only provided valuable resources and training, but also helps to foster a community of practice amongst Welsh educators committed to advancing digital literacy and cybersecurity education. By facilitating direct engagement between teachers and cybersecurity education initiatives such as CyberFirst, educators can be supported to credibly upskill their cybersecurity knowledge while drawing upon supportive relationships from beyond their institutions, promoting scalable and sustainable approaches to cybersecurity education in schools.

3.3 Case Study 3: Pathways Event

In late 2023, the University of South Wales' Newport Campus held a Pathways event, drawing over three hundred young people from across South Wales. This initiative was designed to provide a comprehensive introduction to the diverse field of cybersecurity, aligning with the Cyber Body of Knowledge (CyBOK) framework. The event's primary goal was to demystify the cybersecurity sector, by exposing young people to distinct areas of knowledge and showcase the variety of career paths available in cybersecurity, helping to challenge the notion that it is a monolithic or inaccessible industry. Young people were able to choose which talks and activities to attend, allowing them to explore their own interests and aspirations. A traffic light system was used to highlight sessions that required no technical knowledge (green) to an understanding of specific, cybersecurity concepts (red).

A key strength of the event was its collaborative nature, bringing together representatives from industry, academia and government. This diverse group of professionals offered young people a unique opportunity to engage directly with potential employers, university representatives and course providers. They were able to attend a variety of activities, talks and informal chats with professionals, with topics covering elements such as human factors, attacks and defences, systems security, as well as software and platform security.

However, young people also reported that the most popular activities were oversubscribed, and they were disappointed about not being able to experience everything. Organisers considered whether the self-driven itineraries meant that unpredictable session attendance would negatively impact the attendee experience, and if pre-registration might be the answer – but logistical issues of attendance and individual timetabling meant that it would be overly challenging to organise. However, when sharing their opinions about the set-up, one young person reported that: "*Being able to go and choose which [sessions] to attend also added to the experience and allowed me to explore the university as well as ask questions about cyber,*" demonstrating the value that young people placed upon choice in such situations.

For industry partners, there was a reflection on the types of events provided and their engagement with young people. One industry partner noted that: *"The more engaging the workshop, the better (obviously). Children struggled slightly with just 'lecturing'."* Evidence on the day, and from the feedback forms, demonstrated that young people were more drawn to the practical sessions, and organisers noted that further support should be given to industry collaborators to translate their expertise into engaging activities.

One of the event's most impactful messages was encapsulated in a quote from the organiser: *"It's just trying to get young people to understand, you don't have to leave Wales now to do these jobs.*" This statement highlights a critical aspect of the event's purpose – showcasing the growing opportunities within Wales itself, countering common perceptions that pursuing a career in cybersecurity necessitates leaving the region – a decision that often disproportionately impacts disadvantaged young people from entering cyber-related careers.

3.4 Case Study 4: Girls Event

In July 2024, Swansea University hosted a CyberFirst Girls careers event, aiming to introduce 75 Year 10 girls to potential career opportunities in cybersecurity. The event was designed to address the underrepresentation of women in STEM and digital fields and fostering a sense of community among young women in these areas. The planning process began with outreach to local schools. Initially, some schools committed but later withdrew last minute, citing various reasons including scheduling conflicts and transportation issues. The setback prompted organisers to conduct a second round of outreach, expanding their invitation to a wider range of schools. The involvement of Technocamps, and their wider network of school relationships proved influential in reaching new schools to attend.

The events activities were carefully curated to provide a comprehensive and authentic overview of cybersecurity skills and career opportunities, developed in collaboration with local institutions and industry. Swansea University's and their wider digital services also played a crucial role in supporting the session. The day began with sessions in cyber law, where young people debated crime scenarios and the ethics of AI. The activity aimed to highlight the multidisciplinary nature of cybersecurity, incorporating legal and ethical consideration, and had previously been run as part of the Pathways event in 2023. A further session was delivered that covered being a woman in a cybersecurity role, followed by a hands-on activity on password security, run as a team challenge. The feedback demonstrated that the combination of personal experiences and practical skills proved to be engaging for young people, particularly regarding careers aspirations, including feedback comments such as *"the ladies from Admiral really inspired me to think about working there in the future"* and *"I really enjoyed the law one. I want to be a lawyer now."*

In the afternoon, a local cyber consultancy conducted a capture the flag activity, which some young people found challenging. Feedback indicated that certain aspects of the activity were overly complex for some participants, while others commented on their enjoyment, with one young person responding: *"The CTF was super fun."* This was then followed by a fact-finding session, where ten women from various computing roles across the university participated in a 'speed dating' style activity, where groups of young people had five minutes to learn as much about the individual as possible. However, some of the women noted that students struggled to ask meaningful, work-related questions, either out of politeness or lack of preparation, but feedback from the young people was consistently positive, remarking on how much they learned from these discussions. The event concluded with a prize-giving ceremony, celebrating participant engagement and achievements throughout the day. At the end of the event, one young person reported on her feelings about attendance, saying: *"It was good to speak to girls taking the same subject as me. I now feel part of a cyber community!"*. This demonstrates, that despite the areas for future development, this approach to engagement reconfirms the potential for targeted career exposure in inspiring young women to consider cybersecurity pathways.

4 Challenges and Opportunities

Above, we have outlined a series of first-hand experiences gained from the delivery of cybersecurity education across schools and colleges as part of the CyberFirst programme in Wales. By examining the challenges and opportunities resulting from these case studies, we aim to provide valuable lessons and practical strategies for others seeking to establish or enhance cybersecurity education programmes in their respective regions. From engaging young people to facilitating teacher training to exploring the role of industry collaborations, this following section offers an exploration of the key factors and strategies that can contribute to the development of impactful, high quality cybersecurity education initiatives for young people.

4.1 Engaging Young People

The drive to develop a diverse cybersecurity workforce has never been more pressing. However, as highlighted above, the challenge does not only lie in meeting industry need, but also in creating educational experiences that engage and inspire young people, particularly those who are currently alienated from exploring opportunities in cybersecurity.

One suggested approach trialled through the experiences outlined above, is the need to explore cybersecurity from the viewpoint of young people themselves. By adopting an approach that places young people's perspectives and interests at the forefront of early engagements, we can begin to develop effective and appealing cybersecurity education opportunities (Saglam et al., 2023). For example, the Lemming Pox activity reimagines a medical scenario from the lens of social media, encouraging young people to use cybersecurity techniques while being couched in the familiar terminology of social media. This framing provides a hands-on interactive activity coupled with real-world relevance for young people, which has proven effective in capturing interest and fostering critical thinking skills.

However, this young person-informed approach is often overlooked in favour of top-down initiatives designed by adults and industry professionals, which can fail to resonate with a younger audience. As demonstrated by the redevelopment of the Trailblazers activity, even small changes in the framing, use of tools and hands-on activities can promote deeper engagement, particularly when it is made relevant to their own experiences as young people. The goal is to use personal understandings and interests as gateways to more nuanced and comprehensive understandings of cybersecurity, encouraging long term engagement and progression within the field (Fisk et al., 2023).

Research demonstrates how young people are aware of cybersecurity within their own socio-cultural sphere (Nicholson et al., 2021), but there can be tendency to lean on popular media to understand wider skills and career pathways where they lack knowledge. Leveraging the excitement generated by the cultural touchstones of "hacker culture" may be vital to develop engagement opportunities for some young people (Fisk et al., 2023; Shires, 2020). However, caution must be exercised to avoid perpetuating stereotypes or limiting the appeal of cybersecurity to a narrow demographic, providing opportunities for those who are not traditionally engaged with the monolithic stereotype of cybersecurity skills and careers. To combat this, a bottom-up approach to designing engaging content is crucial and has been well demonstrated as a method of engagement in wider computing education literature (Saglam et al., 2023). For example, using scenarios and challenges that are relevant to young people – such as popular artists, social media and non-technical opportunities with cybersecurity, to help engage those who are dissuaded by the dominant narrative of cybersecurity. Further work should be undertaken to understand their interests and perspectives, while also working alongside educators who possess strong pedagogical knowledge of how to translate these interests into viable, impactful educational engagements.

The regional nature of programmes like CyberFirst offers a promising avenue to address the challenges of exclusionary cybersecurity skill development and career pathways. For example, by focusing on local opportunities and partnerships, these initiatives can provide clearer pathways for young people to get involved in cybersecurity without having to move away from their current support networks – a vital part in encouraging disadvantaged young people to explore with education and career opportunities (Hodgson & Spours, 2013). This was further supported in sessions such as the CyberFirst Girls event, where young people noted the importance of direct engagement with women from local businesses, as well as experiences of wider areas of cybersecurity such as law and ethics, to help them consider realistic pathways into the cybersecurity field.

While these types of sessions demonstrate immediate positive impact on engagement and aspiration development, a significant challenge in evaluating the success of cybersecurity education initiatives is in their long-term nature. The true impact of programmes like CyberFirst can only be fully assessed once young people begin entering the workforce. This underscores the need for longitudinal studies that seek to explore how young people are influenced to pursue futures in cybersecurity, their relevant aspirational and skill development, as well as their pathways into specialised education and careers. This can begin to contribute a comprehensive measure of programme impact, as well as an understanding how to address the challenge of diversity within the field.

4.2 Delivery in Context

The implementation of programmes for cybersecurity education in school settings and contexts presents a unique set of challenges. The following section examines the complexities of delivering educational engagements for schools, highlighting the limitations faced by educators and opportunities that may be able to mitigate or overcome such obstacles. One of the primary challenges in delivering cybersecurity education is the limited expertise of educators in a rapidly evolving field (Saglam et al., 2023). Teachers, already burdened with curricula development, delivery and administrative responsibilities often lack the time, resources and support to develop proficiency in cybersecurity. The knowledge gap can lead to reticence to incorporate cybersecurity topics into their teaching, potentially limiting young people's wider exposure to the field. To address such a challenge, it is crucial to develop materials and support systems that are considerate of the teaching context. This involves creating resources that are not only accurate and up to date, but also pedagogically sound and can integrate easily into the structure of school life.

However, even with high-quality resource, we must consider the challenges of delivery. CPD that is understanding of these constraints and limitations can be effective opportunities to upskill teachers and encourage participation with the CyberFirst network. Delivering these in a lesson-like format, can help to encourage adoption of resources and provide authentic opportunities to upskill. These networks of support are an additional way to provide expertise to teachers and young people, and the importance of building rapport through existing networks cannot be overstated.

Furthermore, access to wider communities of practice and support, such as CSE Connect can be valuable, but it is important to recognise that non-practitioner focused events can be difficult to attend, due to financial and practical limitations, as well as concern about relevance to school practice. In cases where schools are struggling to engage with the cybersecurity education community, further work must be undertaken to understand how to provide inclusive and accessible communities for schools, in order to provide improved pathways to participation for their young people (Fisk et al., 2023) – this can include new approaches to educational delivery, teacher CPD, support to attend networking events that are sensitive to the constraints of the school context, and their digital infrastructure. Working in partnership with these teachers can help formally crystalise these challenges, and present viable strategies to address these limitations to participation in the wider cybersecurity education ecosystem.

Furthermore, understanding the curricular imposition on educational delivery can help drive engagement with schools. For example, the restructuring of the Trailblazer programme into three workshops, has helped align the CyberFirst Programme with the broader goals of the Curriculum for Wales. By incorporating elements of social media analysis, digital literacy, creative technology use, language and problem-solving, the workshops provide a holistic approach to cybersecurity education, with feedback from young people indicating high levels of engagement and learning. Moreover, the cross-curricular design of these workshops allows for greater integration into various areas of the school curriculum, potentially reaching a broader audience of young people and demonstrating the pervasive nature of cybersecurity in different academic disciplines.

By acknowledging the constraints faced by educators in schools, and how this impacts the delivery of cybersecurity education in context, we can continue to reflect and grow the impact of cybersecurity education initiatives. The key remains in a flexible, collaborative approach between stakeholders in schools, industry and academia to help develop an impactful cybersecurity education ecosystem – where limitations and constraints are acknowledged, and efforts are undertaken to meaningfully support all stakeholders – ultimately contributing towards a stronger, more resilient cybersecurity workforce of the future.

4.3 Industry Collaborations

The integration of industry collaborations in cybersecurity education has emerged as a pivotal strategy in preparing the next generation of cybersecurity professionals (UK Government, 2021). These partnerships can offer unique, and authentic perspectives on career pathways and skill development, but present their own distinct challenges regarding how these engagements can be successfully translated to the educational context in ways that are sustainable and scalable.

One of the most important aspects of these collaborations is their role in demystifying career pathways in cybersecurity. By providing young people with direct encounters with cybersecurity professionals, learning about their career trajectories and day-to-day experiences, these engagements can help stereotypes about the people who choose to enter the cybersecurity sector. This exposure can be instrumental in breaking down perceived barriers to entry (Mountrouidou et al., 2019), making the field more accessible and appealing to a diverse range of young people. However, resource constraints present a significant obstacle towards industry engagement, particularly for smaller firms that may struggle to allocate time and resources towards non-business critical engagements. This raises question about how to ensure equitable participation across industry, a potential need for support and incentivisation to encourage a wider range of businesses, and perspectives, to contribute towards the support of such cybersecurity education initiatives.

Furthermore, industrial collaboration can offer the opportunity for authentic, hands-on experiences that can bring the curriculum to life. Through simulations, challenges and hands-on experiences, such as the Pathways event held by the University of South Wales, young people can engage with practical cybersecurity scenarios, bridging the critical gap between theoretical knowledge and practical applications. Direct engagements with cybersecurity professionals, such as the CyberFirst Girls event, demonstrate how industry partnerships can help demonstrate new career pathways and inspire a sense of community. These experiences can help inspire and engage young people, but also provide a tangible sense of what a career in cybersecurity really entails. By combining knowledge exposure with direct industry engagement, the Pathways event served as a powerful tool for inspiring and guiding Welsh young people towards cybersecurity – not only broadening their understanding of the field, but also highlighting the potential for building successful careers in their local communities.

However, the alignment of expectations between educational institutions and industrial partners can present challenges. For example, while schools may seek long-term engagements, companies may prefer short-term, high impact interventions that align with their business priorities. Bridging this gap requires ongoing communication, and CyberFirst Wales has demonstrated how providing appropriate networking infrastructure can support the needs of each party in the relationship – with flexible models of engagement but an ongoing commitment to support a school partner.

Building upon the challenges and opportunities previously discussed, it is also important to highlight the potential disconnect between industry-developed resources and activities, and the pedagogical needs of young people. While industry partners are able to bring their real-world expertise to the table, it may be a struggle to appropriately translate these to an educational context. This disconnect often stems from a misalignment between industry expectation of young people's skills and knowledge, resulting in materials or activities that are too advanced, too technical, or simply not engaging enough for young people. This underscores a need to adopt bottom-up approaches in resource development, with young people and pedagogical experts involved in the creation of learning materials, ensuring that resources are both technically relevant but also pedagogically sound and age-appropriate (Saglam et al., 2023). Such a collaborative approach can help bridge the gap between industry and education, resulting in developing a more nuanced understanding of cybersecurity education best practices.

In summary, while collaborations between education and industry in cybersecurity present a range of challenges, their potential to enrich learning experiences and clarifying career pathways is significant. By adopting strategic approaches and leveraging such opportunities, the CyberFirst programme in Wales can continue to strengthen school-industry partnerships, ultimately contributing to a robust and diverse cybersecurity workforce for the future.

5 Conclusion

While cybersecurity demand outstrips the supply, high-quality cybersecurity education will remain a priority. However, despite high quality work being undertaken in this area, there are still many challenges around engaging young people to develop these skills and consider relevant educational and career pathways. Misconceptions and stereotypes of cybersecurity engage some young people, but drive many away, especially from choices that help them explore the potential of cybersecurity education, and cohesive, high quality cybersecurity education is challenging to deliver in schools due to rapid changes in the field, limited resource and lack of specialist subject knowledge.

Through exploring the experiences of a pan-Wales cybersecurity education initiative through the lens of CyberFirst, as supported by Technocamps, we outline emergent areas of challenge and opportunity towards the betterment of the cybersecurity education ecosystem in the UK. While the reflective experiences presented in this chapter cannot presume to be representative of the wider sphere of cybersecurity education, they serve as a call to action, highlighting the challenges and opportunities needed to improve current approaches to cybersecurity education. This work advocates for continued pedagogical innovation to address the engagement and learning needs of young people in cybersecurity. It also emphasises the critical role of accessible, continuous professional development to ensure a sustainable and scalable approach to cybersecurity education. Finally, it provides a critical perspective on the role of industry collaboration, including current limitation and opportunities, as a way to engage young people who often feel left behind in the narrative of cybersecurity education and careers. The insights gained through this initiative underscore the importance of reconceptualising collaboration and support between industry, academia and education sectors to provide an impactful, sustainable and scalable cybersecurity education ecosystem.

References

Cabinet Office, National security and intelligence, HM Treasury, & The Rt Hon Philip Hammond. National Cyber Security Strategy 2016 to 2021 (2016)

Chase, J.D., Uppuluri, P.: High school cybersecurity? Challenge accepted–Radford University's RUSecure CTF contest for high school students. J. Colloq. Inf. Syst. Secur. Educ. **9**(1) (2022)

Coutinho, S., et al.: Cyber security skills in the UK labour market 2023 (2023)

Department for Education. National curriculum in England: computing programmes of study. National Curriculum in England (2013)

Department for Science, Innovation & Technology. Inspiring and equipping future talent: scaling impact of the CyberFirst programme. UK Government, 24 May 2024

Fisk, N., Kelly, N.M., Liebrock, L.: Cybersecurity communities of practice: strategies for creating gateways to participation. Comput. Secur. **132**, 103188 (2023). https://doi.org/10.1016/j.cose.2023.103188

Furnell, S.: The cybersecurity workforce and skills. Comput. Secur. **100**, 102080 (2021). https://doi.org/10.1016/j.cose.2020.102080

Hodgson, A., Spours, K.: Tackling the crisis facing young people: building "high opportunity progression eco-systems. Oxford Rev. Educ. **39**(2), 211–228 (2013).https://doi.org/10.1080/03054985.2013.787923

Kemp, P., Wong, B., Hamer, J., Copsey-Blake, M.: The future of computing education considerations for policy, curriculum and practice (2024)

Mountrouidou, X., et al.: Securing the human: a review of literature on broadening diversity in cybersecurity education. In: Annual Conference on Innovation and Technology in Computer Science Education, ITiCSE (2019). https://doi.org/10.1145/3344429.3372507

Neutens, T., Wyffels, F.: Teaching computing in primary school: create or fix? In: ACM International Conference Proceeding Series (2018). https://doi.org/10.1145/3265757.3265791

Nicholson, J., Terry, J., Beckett, H., Kumar, P.: Understanding young people's experiences of cybersecurity. In: ACM International Conference Proceeding Series (2021). https://doi.org/10.1145/3481357.3481520

Ofsted. Research review series: computing (2022)

Saglam, R.B., Miller, V., Franqueira, V.N.L.: A systematic literature review on cyber security education for children. IEEE Trans. Educ. **66**(3), 274–286 (2023).https://doi.org/10.1109/TE.2022.3231019

Shires, J.: Cyber-noir: cybersecurity and popular culture. Contemp. Secur. Policy **41**(1), 82–107 (2020). https://doi.org/10.1080/13523260.2019.1670006

The Royal Society. After the reboot: Computing education in UK schools (2017)

UK Government. Department of Digital, Media, Culture & Sport. Department of Science, Innovation & Technology. CyberFirst Evaluation (2021)

Welsh Government. Cyber action plan for Wales (2023)

Authentication in the Connected World: Perceived Password Security Is Inflated

Drew Bassett(✉), Nicholas Micallef, and Berndt Müller

Department of Computer Science,Swansea University, Wales, UK
{2405694,nicholas.micallef,berndt.muller}@swansea.ac.uk

Abstract. Passwords are still a popular authentication approach, even as biometric authentication has become a widespread method to access our phones. Although password managing tools can aid users in storing and securing passwords, many are unaware of their existence. In our research we adopt an interpretative research methodology in which we conduct semi-structured interviews with individuals in a UK law firm and we compare their responses to guidelines offered by NIST on passwords to assess how secure their passwords are in comparison to passwords generated using an application developed for this research project. Overall results vastly differed from the NIST guidelines, and it was established that participants are unaware of the concept of external individuals attempting to gain their passwords and are instead attempting to protect them from their peers. This research suggests that our participants should receive additional cyber security training to bridge the gap between perceived and actual security regarding password complexity and to acquire the skills required to start using modern password managers.

1 Introduction

A survey in 2023 showed that although over 50% of respondents were 'excited about passwordless options' for authentication, almost a third were not and preferred using 'something that you know', such as a PIN or password, for authentication [10]. Further analysis in this survey shows that almost 60% of respondents memorised their passwords, 34% used pen and paper, 23% a document on a computer, and only 30% facilitated a password manager. The majority of passwords, according to the survey, were names of family and friends, lyrics, or other well-known names (cities, landmarks, or celebrities). The 2024 edition of the survey [11] shows almost no change in users' authentication behaviour. The question of reuse of passwords across accounts has not been reported on in the most recent survey results, but judging from earlier surveys, many users share passwords across several accounts. This indicates that despite popular media reporting 2024 as the year of a 'workforce free from passwords' [36], a significant proportion of users and sites still rely on authentication by password.

P. Legg et al. (Eds.): CSE 2024, LNNS 1213, pp. 170–191, 2024.
https://doi.org/10.1007/978-3-031-77524-6_10

While the latest UK National Cyber Strategy [29] makes a distinction between the real world and the virtual world that we enter when "we go online to communicate, work and conduct everyday tasks", the reality is rather more complicated in that the real and virtual worlds are profoundly intertwined. Arguably, the 'virtual world' has become part of the real world for a large proportion of humans [31]. We call this the connected world. A particular feature of the connected world is the fact that some entities exist in both parts of the connected world, e.g. a password is just a sequence of symbols that can extend its existence across the real and virtual worlds. It can exist on a sheet of paper as well as in digital form.

Entities in the connected world are usually associated with a set of identifiers, some of which are used for the purpose of authentication and make them subject to monitoring and surveillance [51,52]. Typically, identification systems serve as the foundation of authentication systems. The latter can be used to ascertain an entity's identity [18] to allow or deny them access to certain parts or aspects of systems. This is the role of an access-control system, that can be hierarchical and/or role based. The entities we usually think of in this context are humans. For these individuals, authentication is granted on a combination of: something you have; something you know; something you are; or something you do [13].

The advent of AI has allowed leaked password lists to be used as a basis to generate potential passwords that users may have legitimately used [55,56]. Combined with regular increases in computing power, it is now easier for hackers to gain access to weakly secured accounts. The use of protections such as multi-factor authentication alone is not enough when websites have been found to disable them to allow account recovery [8].

Ensuring the integrity of identification systems is essential for the authentication process, and numerous guidelines are available to provide instructions on how to achieve this. One such digital identity guideline is provided by the National Institute of Standards and Technology (NIST) [25] who are a part of the US Department of Commerce. The NIST guidelines are highly regarded among security professionals worldwide. A key aspect of this set of guidelines is the instruction of requirements for knowledge-based authentication methods, i.e. passwords.

Despite extensive research on password usage across diverse communities, our analysis of previous studies revealed no research specifically investigating the adoption of NIST password guidelines among participants in the legal sector when setting passwords. To address this gap our research seeks to answer the following research question:

RQ: When presented with passwords generated using NIST best practices, are users in a UK law firm likely to generate passwords of similar strength?

To answer this question, we investigated if users in a UK law firm worked in accordance with the NIST password best practices. Semi-structured interviews were performed with a group of participants from a UK law firm. The interviews asked a range of questions regarding the participants' own passwords. Subsequently, participants were asked to reflect upon passwords created using the NIST standards. It was observed that participants initially had an overestimated perception of their own password strength, which diminished as the interviews progressed and they recognised the criteria for constructing a secure password according to industry standards.

This paper contributes to the literature on human behaviour in authentication and cyber security through qualitative research conducted in a niche population, a UK law firm, which to our knowledge, has not been previously studied. Additionally, our findings contribute to the field of cyber security education by highlighting that the two password training modules provided to the participants contain outdated recommendations.

The paper continues by providing a background on password authentication methods and the human factors that influence them, followed by describing various governmental-published guidelines aimed at strengthening authentication. It then discusses related work, explains the research method, presents and discusses the findings. We conclude the paper by suggesting potential future work and summarising our key contributions.

2 Background

The persistent use of passwords for authentication makes them a prime target for attackers aiming to gain access to user accounts through methods such as brute force attacks, credential stuffing, phishing, and shoulder surfing. In this section, we explore recent attacks against passwords, introduce guidance published by government agencies for enhancing user password security, and describe alternative methods that the research community has studied as an attempt to move away from traditional text-based passwords.

2.1 Recent Cyber Attacks on Passwords

In 2024, WIRED reported[1] a cyber attack on Snowflake, a cloud storage company which has many large clients such as Ticketmaster and Santander. It seems that this attack resulted from malware installed on a computer of one of Snowflake's contractors, which allowed discovery of plain-text usernames and passwords. In addition, previously-harvested credentials stored in leaked datasets were used to perform a brute force attack.

Upon discovery of a breach, companies have a legal (through GDPR) and ethical responsibility to notify their customers. While it is unclear which specific previous breach the passwords used in this attack were harvested from (e.g. RockYou, Yahoo, LinkedIn, etc.), it is likely the owner of the account was notified of their disclosure by the breached service. Storing credentials in an unencrypted format greatly increases the risk of information disclosure, especially high-value accounts such as that of the accused contractor.

[1] https://www.wired.com/story/epam-snowflake-ticketmaster-breach-shinyhunters/.

In 2024, Roku, a streaming service, experienced two breaches attributed to credential stuffing,[2] caused by the reuse of previously harvested credentials, resulting in over half a million compromised user accounts.

Passwords continue to be used due to their ease of implementation, while attacks targeting passwords continue due to their widespread adoption over the internet. The research community as well as government bodies are actively engaged in providing safeguards to protect users against these ongoing threats.

2.2 National Efforts and Security Standards

Previous guidance from the UK government, issued by the Communications-Electronics Security Group (CESG) [16], recommended using four random dictionary words, which was expected to result in longer, more secure passwords. The UK's National Cyber Security Centre (NCSC), which replaced CESG, now recommends using three random words for password creation rather than using complex passwords. Their reasoning being that three random words typically result in a long password that is easier for users to remember and more likely to be a unique combination not previously used, therefore does not exist in a blacklist [40].

The NCSC publishes various guidelines in their Cyber Essentials scheme, a certification that businesses can pursue to secure their practices. Cyber Essentials [41] advises using passwords with a minimum length of twelve characters or a minimum length of eight characters, provided the password is checked automatically against common blacklists.

In contrast, the British Standards Institute (BSI), the national standards body who is independent from the UK government, defines a strong password in ISO 29115:2013 [13] as one that includes non-dictionary words and includes a mix of capitalisation, numbers, and special characters.

The inconsistency between these guidelines results in confusion regarding security best practice. Interestingly, ISO 29115:2013 [13] references the NIST guidelines [25], which are the US equivalent to the NCSC guidelines. Next, we describe these guidelines in more detail.

2.3 NIST Digital Identity Guidelines

The NIST guidelines [25] regarding passwords state the following:

- user-generated passwords are to be at least eight characters long.
- passwords checked against a list of compromised passwords are not to be used if the password-requiring system is to check against the blacklist.
- enforcing a mix of uppercase letters, symbols, and numbers provides limited benefit, but password length plays a key part in password strength.

[2] https://edition.cnn.com/2024/04/12/business/roku-security-breach-user-accounts/index.html.

The guidelines further suggest, but do not endorse, enforcing the use of different characters in passwords, including all printable ASCII and Unicode characters. This reduces to 95 printable ASCII characters available on a standard English keyboard. ASCII is a subset of Unicode, which contains symbols from other languages that are not typically available on a standard English keyboard and thus may not be feasible for users to include in their passwords.

The requirement for passwords to be at least eight characters long and checked against a blacklist somewhat aligns with the guidance provided by NCSC in Cyber Essentials [41], but not their recommendation of using three random words [40]. NIST does not stipulate that all passwords be checked against a blacklist, only that passwords found on blacklists must not be used if a system does implement a blacklist. However, NCSC Cyber Essentials [41] requires the use of blacklists if password policies recommend a minimum length of eight characters. The BSI's ISO/IEC 29115:2013 [13] does not mention blacklists, instead it recommends the inclusion of complexity requirements, which NIST and NCSC directly advise against.

2.4 Alternatives to Traditional Passwords

The research community has explored several authentication methods aiming to develop approaches that could eventually replace passwords. Alternatives to text-based passwords include graphical passwords, which have been found to be just as usable [42] and memorable [45] as traditional passwords, though they have yet to be widely adopted by the industry.

One way to mitigate the risk of weak passwords is through multi-factor authentication. Abbott and Patil [1] found that enforcing the use of multi-factor authentication on all university systems was highly detrimental to the user experience, and that 92% of participants preferred its use on some systems rather than all.

In contrast, biometric technology appears to be a preferable method of multi-factor authentication due to its security and ease of use [28]. Many users opt to use biometric authentication on their mobile devices instead of passcodes and patterns [50], but deploying biometric authentication on other devices can be expensive [2]. A study on user acceptance of biometric technology found that privacy concerns significantly influenced the adoption of the technology [48]. As long as users have the option to sign-in only with a username and password, there will always be the need to protect their security and privacy.

In the next section, we discuss studies on how users perceive protecting their online assets and what behaviours they exhibit when using passwords.

3 Related Work

Most relevant to our work, a study of over 20,000 websites found that outdated NIST guidelines were being followed: 42.1% of sites adhered to the 2004 guidelines, 16.7% adhered to the 1985 guidelines, and only 30.8% adhered to the latest NIST guidelines [5]. The older NIST guidelines have shorter minimum length requirements, allowing users to choose weaker passwords on many websites. Another study of 120 popular websites found that 75% do not prevent users from selecting the most common passwords [33]. Users cannot rely on website security policies and must be aware of and willing to use current best practices.

In this section, we discuss previous studies on password policies and the issues affecting user motivation to adhere by them. We focus on the most recent work due to the ever-changing technology landscape, but we acknowledge plenty of earlier research on this topic and recommend readers to refer to this literature [3, 12,19,21,35,38,47].

3.1 Password Policies

Administrators use guidelines such as those offered by NIST and their national equivalents to create password policies for their services and organisations. Gerlitz et al. [23,24] conducted two studies on German workplace password policies from the perspective of system administrators. In their 2019 study [23], they found that 89% of company policies required to be at least eight characters long, aligning with the NIST requirements. However, 89% also imposed complexity requirements, which contradict NIST's guidelines but comply with their German equivalent, BSI. The 2023 study [24] analysed password policies over a three-year period and found that the number of required character classes decreased throughout the study but remained at 76% in the final survey held in 2023. Character classes include uppercase letters, lowercase letters, special characters, numbers, and symbols such as Chinese language characters [24].

It was reported that 92% of password policies are enforced through technical methods [23], although it is unclear if these policies apply only to computer and email accounts, or extend to websites as well. The authors did not question participants about the technical methods, though it could include techniques such as implementing complexity requirements through a Windows Group Policy. Alroomi and Li [5] automated a search of over 20,000 websites and found 12% lacked minimum length requirements and 30% did not support special characters, and Gautam et al. [22] found only 39% of studied websites met NIST's minimum password length requirement. While companies may impose restrictions on some accounts, others may be created by users where the organisation cannot technically deploy password restrictions, leaving the responsibility on the user to create a secure password.

Whilst password policies provide requirements, user motivation to adhere to these policies is a different matter. Not every policy can be enforced through technical measures, so it is critical for policymakers to understand user motivations towards account security to address any related shortcomings.

3.2 Protection Motivation Theory

The theory that explains why people are motivated to protect assets is conceptualised in Rogers' work [46] on Protection Motivation Theory (PMT), which models how an individual perceives the severity of a threat and their efficacy in engaging in preventative behaviour to protect themselves.

Humaidi and Alghazo [30] developed a model using PMT to investigate how users are motivated to generate secure passwords, secure their devices, and exhibit proactive behaviour. Although the participants were recruited through snowball sampling, which may induce bias, the demographics indicate respondents from various positions in diverse industries, allowing for some generalisability. The findings showed that participants were not influenced to create secure passwords based on their perceived severity of the threat, but were positively influenced by their perceived susceptibility and self-efficacy. In other words, the negative consequences of weak passwords did not significantly affect their behaviour in creating stronger ones. Instead, the recognition of a potential threat to their passwords and their perceived ability to create a secure password motivated them positively. Research into Facebook users also found similar results regarding threat severity and perceived susceptibility, but not self-efficacy, which was determined to not impact motivation [32].

In a related study evaluating PMT among teenagers [39], researchers instead found that perceived severity had a weak effect on motivation, and perceived vulnerability and self-efficacy had a weaker effect on motivation than in Humaidi and Alghazo's study [30]. While methodological differences may account for these variations, the contrasting results suggest that further research is required to discover what motivates individuals to use passwords, how they choose strong passwords, and when they decide to use them.

A user's motivation to create and use a strong password is intrinsically related to their ability to remember their chosen password. The ability to memorise passwords relates to a user's self-efficacy, but is also an active research topic in its own right.

3.3 Memorising Passwords

In a study examining memory anxiety and password behaviour [53], it was found that 98% of participants either reused the same password or made slight modifications and reused it. The primary reasons provided were the ease of remembering a single password and the burden of managing numerous passwords. The study identified that metamemory anxiety significantly predicted both password reuse and modification, indicating that users tended to use the same password due to anxiety about forgetting passwords if they were required to use different ones. The study highlighted that anxiety was the only factor influencing password behaviour [53], while factors such as knowledge of password memory strategies, self-belief in remembering passwords, and personal motivation to remember them did not significantly affect actual behaviour.

Although the study did not mention password managers to participants, previous research acknowledges user reluctance to adopt them [53]. Anxiety also emerged as a significant barrier to adopt password managers in Mayer et al.'s study [34], where respondents expressed concerns about the security of password managers and the risk of storing all their passwords into one system.

Many users unwillingly utilise password managers integrated into web browsers [4], though these lack the portability of standalone password managers. Recent studies have revealed that users often overlook the password generation feature of their password managers, primarily using them for convenience rather than for enhanced security [4,7].

Innovative approaches to address memorability issues include the use of gamification. For instance, the Password Game[3], gained viral popularity in 2023, although its success stemmed more from its unrealistic requirements than its actual effectiveness in enhancing memorability. Alternatively, researchers employ techniques grounded in psychology to leverage imagery and memory cues to improve password recall [37]. Such gamification serves as a nudging mechanism to positively influence user behaviour.

3.4 Improving Password Security Through Nudges

Nudging user decisions is a contentious issue, with ongoing research into the use of dark patterns by websites to influence users into sharing personal information or making unintended purchases [26]. Regardless, Nudge Theory was introduced as a mechanism to encourage users to make better choices [49]. A classic application is through password meters, where recent literature reviews have shown that the presence of password meters does not adversely affect password memorability. Yet, the effectiveness of different types of password meters remains inconclusive in the literature [57].

Subsequently, Zimmermann et al. [57] propose their own password meters, including one that indicates how long it would be before the user would have to change the password, which contrasts with NIST guidelines [25] that advise against forced password expiry.

Amador et al. [6] explored the use of nudging and interactive prompts, finding that negatively framed nudges highlighting potential financial loss and identity theft had the greatest impact on improving password strength. This contrasts with findings that threat severity, as studied by Humaidi and Alghazo[30] and Kusyanti [32], did not significantly influence motivation.

Other studies have explored personalising nudges to allow greater influence over users and by directly targeting their own motivations. Guo et al. [27] propose a dynamic personalised password policy where user personality traits are inferred from their initial password choice when creating a new account, and a personalised nudge is then displayed based on these traits. However, the effectiveness of this approach was limited by the small sample size in their study. In contrast, Qu et al. [44] determine user personalities by first having a much larger group of participants fill out a questionnaire to assess their General Decision-Making Style (GDMS). This determines which type of nudge is assigned to each participant. This study found that all types of nudges had significant effects, with social-negative nudges having the strongest impact.

[3] https://neal.fun/password-game/.

A more extensive study by Peer et al. [43] asked participants to complete a psychometric questionnaire, including GDMS questions, before creating an account where they encountered a personalised nudge. Similar to previous studies, personalised nudges were found to significantly influence users to create stronger passwords. Notably, while the study highlighted the effectiveness of personalised nudges, it also revealed that a standard password meter, i.e. without personalisation, was effective in both the randomly assigned and personalisation conditions. The impact of personalised nudges appears to be particularly evident in scenarios where users faced a steep decline in password strength when subjected to alternative nudges focused on time required to crack passwords.

4 Method

Our research employed an interpretivist approach to uncover the perceptions of our participants. Given that participants were required to create passwords for accessing various workplace systems, and likely managed their own passwords for personal systems as well, we aimed to explore their rationales behind these practices. This section outlines the method used to conduct our research.

4.1 Procedure

We conducted semi-structured interviews (**n=12**) where participants were asked a series of questions regarding their personal password habits, strategies for creating passwords, and their perceptions on passwords created specifically for our study. The interview questions are listed in Table 1, although additional questions were asked as needed.

Questions six and seven were asked separately for each of the passwords presented to the participants, which were:

- Bro@w2nSugarEgg
- *GrassBigRemembe7r
- WrongOccurSurp1ri,se

The participants were presented with the passwords on flashcards during the interviews to ensure they could focus fully on analysing them, rather than relying on their memory if the passwords were spoken. Figure 1 demonstrates how passwords were displayed to the participants.

These passwords were generated using a password generator research instrument coded in Python specifically for this study. The generator adhered to the NIST guidelines described earlier and used a list of English words sourced from the 1,000 most frequent English words present in a public domain book [15].

The script concatenated three random words and added an additional random word if the combined length of the first three words was less than 8 characters. The choice of using three words was made to ensure the passwords met length requirements, guided by NCSC recommendations [40]. While the NIST guidelines discourage complexity requirements for user-generated passwords, we

Table 1. Interview Questions

No.	Question
1	Do you think your own passwords are secure?
2	Do you tend to forget your passwords?
3	When creating a new password, you are told it is not secure and you need to strengthen it, how would you improve it?
4	What do you think makes a password secure?
5	Have you ever used a password generating tool? Why? (If yes - Did you find their passwords easy to remember?)
6	Consider your most secure password. Do you think the following is more or less secure? Why?
7	Do you think you would forget this password?
8	What do you think is the security recommendation for password length?
9	Do you think password length is more or less important than including symbols and numbers. Why?

included some complexity in the passwords based on our understanding from prior research.

Participants were explicitly instructed, as part of the consent process, not to disclose their own passwords or hint at them, but only to reflect on the presented passwords. This was critical to maintain the security of their accounts. Additionally, participants were asked not to use the presented passwords in their own accounts. Participants provided explicit consent before the interview and were provided with an information sheet to retain after the interview.

Fig. 1. Passwords, such as the one displayed here, were presented individually on flash-cards.

We did not specify whether participants should focus on their work-related passwords or personal passwords; rather they were asked to reflect on any passwords they create. This approach was informed by findings from Brockbanks and Butler [14], who reported that up to 56% of users reuse passwords across both work and private accounts.

The interviews were conducted without recording equipment, and responses were transcribed directly by the lead researcher into a text document during each interview to ensure accuracy. Prior to the main study, the interview questions were piloted with one participant (**n=1**) to refine coherence and ensure the questions elicited useful responses. The pilot participant's responses were not recorded and were not included in the main study.

4.2 Recruitment

Participants were recruited from a UK law firm and engaged in face-to-face semi-structured interviews. Staff members represented various roles within the firm, ensuring a broad range of responses. Educational backgrounds ranged from high school education to advanced degrees such as master's degrees. Participants also varied in terms of their tenure within the law firm and their overall work experience since completing their education. Interviews were conducted over a three-week period, providing a cross-sectional view over time. Each interview was held in an informal setting to create a comfortable environment, encouraging participants to provide truthful responses on a topic that can feel personal and intrusive. Recruitment was managed by the lead researcher, which may introduce bias, although participants were selected irrespective of their educational background or other demographics, as long as they were employed by the law firm.

In the UK, law firms can handle legal aid cases if they meet specific accreditations that demonstrate high standards of legal practice management. One such accreditation, Lexcel, is awarded by the Law Society, and it requires law firm to hold Cyber Essentials certification. Cyber Essentials certifications can be either standard and plus; Lexcel accepts either variant. As previously described, these certifications impose password restrictions similar to the NIST guidelines [25,41]. Therefore it was expected that participants would adhere to password requirements aligned with NIST guidelines, as required by their firm's policies.

4.3 Organisation Policies

Participants have undergone various levels of cyber security training as required by their employment in the law firm. This training consists of four cyber security modules covering topics such as phishing, software updates, and a dedicated module on passwords. Upon joining the firm, participants complete all four modules, and subsequently, every year they get enrolled in a refresher course on one of these four modules.

The password module specifically requires passwords to be a minimum of 8 characters long and to include a mixture of numbers, symbols, lowercase, and uppercase letters. This guidance aligns with the NIST standards [25] regarding password length, although the firm imposes additional complexity requirements that are not in line with the guidelines.

Participants may have received additional cyber security training in previous roles, which could influence their responses. However, the direct connection between their prior training, which may have occurred some time ago, and their current responses cannot be definitively established.

According to the organisation's password policy, passwords must meet the minimum length requirement of 8 characters, and include a least one uppercase letter and one number.

Whilst the NIST guidelines [25] advise against complexity requirements due to predictability, acknowledging that highly complex passwords are more likely to be written down, the complexity requirements enforced by the organisation may stem from recommendations by Microsoft[4], where such requirements are optional for administrators when configuring password group policies.

4.4 Ethics

Our study was approved by the university's Institutional Review Board (IRB), which reviewed and approved both the ethical considerations and the interview questions used in the research. The participants in our study were consenting adults employed at the law firm, and none of them were considered vulnerable individuals. Each participant was instructed to provide truthful responses without attempting to anticipate what the researcher might want to hear.

To ensure confidentiality and avoid any potential repercussions from our findings, we have refrained from disclosing specific details about the law firm, its training providers, or any commercially restricted resources that are copyrighted. Our analysis of these resources was based only on discussions with our participants.

4.5 Limitations

The small sample size of the interviewees restricts our contributions to the specific law firm. Although, we are confident that some of the contributions could easily hold for other organisations in the same sector, especially those who use similar training resources. Furthermore, this research offers insights into password practices at a single point in time; the findings may evolve over the years due to changes in workplace policies or increased awareness of security issues among participants. Although this factor did not affect our research question, future work should consider defining research questions that study change of password security perceptions over time.

Several participants commented on the length of the presented passwords, which might have influenced their responses to the questions that probed them about their ability to recall them. To address this limitation the study could have benefited by the use of generated-passwords composed by only two words, with

---

[4] https://learn.microsoft.com/en-us/previous-versions/windows/it-pro/windows-10/security/threat-protection/security-policy-settings/password-must-meet-complexity-requirements.

a third word only added to meet the minimum password length requirements. This factor should be taken into consideration when designing future studies.

5 Findings

Upon completion of the interviews, the responses were analysed by the lead researcher using the transcribed answers which were written during the interviews. The responses were hand-coded, revealing common themes in participants' responses towards password security. A deductive analysis approach was taken, with the expectation that responses would align with predefined categories based on the interview questions.

Table 2 presents an analysis of the results, including the main quotes extracted from the participants responses. Each theme identified is compared against the recommendations in the NIST guidelines [25]. This comparison highlights the alignment or divergence between participants' thoughts and behaviours and the recommended best practices.

More specifically, the security of participants' passwords varied. Initially, the largest proportion of participants perceived their own passwords to be secure. However, most participants could not remember their passwords and some admitted to regularly reusing passwords. While participants initially thought their passwords were strong, deeper probing into their password creation strategies and comparisons with examples of strong passwords revealed that their passwords were not as secure as they first believed.

Regarding the recommended length of passwords, no participant suggested a length shorter than 8 characters. Three participants recommended 8 characters, two participants recommended 12 characters, and three suggested 15 characters. The remaining responses were within the 8–15 characters range. Only one participant mentioned a maximum length, specifying a range of 10–12 characters.

The presented passwords were generally perceived to be more secure than participants' own passwords, though participants expressed concerns about their ability to remember them. Some identified phrases in the passwords that could help their memory. A few participants reported that certain passwords lacked security due to not including sufficient symbols or numbers, despite each password including at least one symbol, one number, and some random uppercase characters. Most participants prioritised symbols and numbers over password length, which is in direct contrast with the NIST guidelines [25], likely influenced by the organisation's policies and training.

To enhance password security, the consensus was that adding symbols or numbers would be effective. This approach diverges from the NIST guidelines [25] but aligns with the law firm's password policy and training, which recommend, but do not require, including such elements for complexity. Most participants had never heard of password generator tools, which could assist with creating strong password. Those who were aware of them, perceived the generated passwords as too complex to use and remember.

The preconception that password generators provide overly complex passwords contrasts with the responses given to the questions related to the

Table 2. Analysis of interviews

Theme	Example Quotes	NIST Guidelines [25]
Security of own passwords	*Do you think your passwords are secure?* "I would like to think so." "No they are practically the same for everything"	Users thought their passwords were secure but were not yet asked to compare them to guidelines.
Memory of passwords	*Do you tend to forget your passwords?* "Yes, always changing them" "Yes" "No ... I use the same one for everything"	No direct mention of using the same password for different services, but if passwords are checked against a blacklist, then passwords present must not be used.
Securing a password	*How would you secure a password?* "random words" "symbols at the end" "hyphen or dash or a forward stroke in the middle"	Adding symbols and numbers provides some benefit to security, but the key factor is password length. No guidance on the location of uppercase letters, symbols, or numbers.
Use of password generation tools	"generated passwords are too complex, and I would never remember them"	No direct mention of generators.
Passwords generated using application	"only has one number in it" "no one will guess this" "too long" "less secure as is more readable" "less secure as there is only a small amount of variation" "[more secure due to the] comma being in weird place" "difficult for someone to guess"	Recommended minimum length of 8 characters, and presented passwords were longer than the minimum. No information on how many uppercase characters, symbols, or numbers should be included, only that there should be a mixture.
Password length	"depends if you remember it"	Minimum of 8 characters and maximum of at least 64 characters.
Length, symbols, or numbers?	"If it is shorter and has more numbers and symbols it is more secure" "they would try the longer ones first" "If the symbols are in random places, they can't guess them"	Length plays a more important part than symbols, numbers, and a mixture of capitalisation.

password-generating research instrument used in this study. It is likely that participants had encountered password generators producing non-human readable passwords, which are difficult to or use without a method to store them, such as a password-manager or written notes. NIST guidelines recommend against enforced complexity requirements, suggesting that user-friendly passwords can still meet security standards. We note that free password manager applications are available, including password generators that adhere to the NIST guidelines of length, avoid enforced complexity, and ensure generated passwords do not exist on blacklists.

Our findings echo the misconceptions observed by Wu et al. [54], where participants believed that five-word passwords were insecure due to the lack of symbols and numbers. Wu et al. emphasised the need for user education on the security of using computer-generated passwords, despite their research indicating a low recall rate when using such passwords. However, in their work, the use of password managers significantly improved the recall rates, leading the researchers to recommended password managers to their participants.

It is evident that our participants created their passwords to prevent individuals who they personally know from guessing their passwords. These attackers could also use personal information to infer their passwords, rather than relying on external hackers to crack their passwords using brute force methods. One participant reported that a secure password should be something personal that others cannot guess, while another participant argued that a password should contain nothing personal. Another participant mentioned avoiding common answers to security questions like addresses or a maiden name. Hackers might employ phishing techniques and social engineering to gather such information, but they could also resort to brute force attacks or creating fake log-in pages to directly obtain passwords.

Moreover, while some participants had an understanding of what constitutes a secure password, they were often unwilling to apply those practices to their own passwords, despite initially believing their passwords to be strong. This reluctance appears to stem from anxiety about forgetting their passwords, as supported by the previously discussed work of Woods and Siponen [53]. It is important to note that mitigations such as password managers exist to help users overcome these memory challenges.

Prior work has investigated potential reasons why password managers are not being adopted. For instance, Amft et al. [7] conducted an analysis of commercial password managers, including both standalone and browser-based options, and found that only half of these tools offered tutorials on how to use them. This lack of educational resources provided by password managers could be a contributing factor to their limited adoption among regular users. Moreover, in a recent study by Dumaru et al. [20], misconceptions were identified regarding the security mechanisms of password manager users. Users reported that passwords might be visible to the password manager provider or external entities. In addition, there seems to be a lack of awareness about the capability of using password managers across multiple devices, which could potentially hinder their adoption if users perceive them as less useful for this reason. However, such limitations could be easily addressed by companies who may have the resources required to offer in-house training for employees to improve their awareness and knowledge about such tools.

The commercialisation of password managers is another plausible reason for their limited uptake, with some negative experiences towards them being attributed to lack of features on free accounts [7]. According to a survey, 66% of respondents indicated they would be willing to use a password manager if their organisation offered it for free [34]. While browser-based password managers are free and are widely used by users [4], their password generation features vary. Some lack robust password generation features, while others require users to log into their browsers, which may deter less tech-savvy users or those hesitant about security implications. These factors contribute to a reduced adoption rate, due to users' comfort level with technology and their organisation's willingness to invest in password management solutions.

Our findings underscore the necessity for cyber security training to address not only the gap between perceived and actual password security, but also convey the skills required to use modern password managers effectively. While browser-based password managers are convenient for users [4,7], they do not enhance security if they are used to store weak passwords. It is critical for users to either create strong passwords themselves or utilise password managers capable of generating secure passwords—a feature currently utilised by only 7% of users [4]. This dual approach would ensure that users adopt robust security practices and leverage tools that safeguard their digital assets effectively.

6 Discussion

As described in the method, the law firm our participants work requires the completion of four cyber security modules, covering topics such as phishing, software updates, and a dedicated module on passwords. The latter module explicitly states that passwords must be a minimum of 8 characters long and include combination of numbers, symbols, lowercase, and uppercase letters. Therefore, the organisation's password policy aligns with the NIST guidelines [25] concerning length but diverges from the guidance regarding complexity requirements.

The information provided in the training appears to be outdated, which is unexpected given the two resource providers come recommended by LawNet, an exclusive membership scheme for law firms that provides recommendations for outsourced partners. A study examining cyber security tutorials on YouTube [17] found a common weakness: the importance of password length was frequently emphasised, but guidance on complexity was often missing from the analysed videos. This oversight underscores the challenge organisations face in ensuring their employees adhere to current cyber security guidance when their training providers are delivering outdated information.

The current training is conducted online and includes reading materials, watching several videos, and various interactive segments, including one where users are asked to rank passwords from strongest to weakest. Our findings indicate that alternative forms of training should be explored, such as workshops and events, as recommended by Mayer et al. [34] for implementing password managers in their organisation. Their research indicates that awareness levels can significantly increase through word-of-mouth. Furthermore, NCSC offers a free online training course[5] covering strong passwords and other cyber security topics. However, they currently do not provide offline resources (other than an infographic summarising the topics covered in the course content) which could be used to facilitate workshops or in-person training sessions.

Despite completing the required training, our findings reveal that participants are not applying what they have learned. The participants inflated self-efficacy may be attributed to their completion of the four training modules, as one might expect that such comprehensive training leads to a high level of knowledge and practice. A large portion of the training is delivered through video format, which Christmann et al. [17] argue that can inadvertently foster an inflated sense of security among viewers. This discrepancy between perceived and actual practice highlights a potential gap in how effectively the training content translates into real-world behaviours.

Although the mandated training may have been outdated, our findings underscore the need for regular and recurring training to reinforce the covered concepts. It is evident from our findings that participants were not strictly adhering to the principles emphasised in their training, highlighting the importance of ongoing education to keep pace with evolving cyber security guidelines.

7 Future Work

Future work could further investigate the employees of the law firm to gather additional insights and determine if the findings from this small sample are representative of the broader population, which consists of around 100–250 employees. Moreover, a follow-up study could be conducted with the same participants at a later time, with revised interview questions to mitigate bias, to investigate if their perceptions evolve over time.

There is also an opportunity for research into the commercially available password training programs designed for employees to assess their adherence to the latest password guidelines. As observed in our work, the two providers used by the law firm do not incorporate the most recent guidelines in their resources.

8 Conclusion

Our research has revealed a gap in the participants' knowledge on who passwords need to be secured from. It is evident that there is the need to increase awareness among participants and their colleagues about the range of potential threats

[5] https://www.ncsc.gov.uk/training/v4/Top+tips/Web+package/content/index.html.

beyond those that exist over their shoulders. Addressing this gap is essential for enhancing cyber security practices within this UK law firm. As for our research question, it appears that users in this UK law firm are unlikely to generate passwords that fully adhere to the NIST guidelines. Thus, we recommend that the organisation revisits its password policy to align with the recommendations from NIST and NCSC. It appears that the current organisation policy may have been developed based on guidance from their training providers. In addition, we recommend that these providers ensure that their content remains updated to reflect the latest cyber security guidelines and best practices. This alignment will help enhance the overall security practices of organisations that adopt their training resources.

If after providing additional training the organisation finds that users resist in adopting secure password practices, alternative options for account security could be made available. These options might include multi-factor authentication or implementation of single sign-on using their work accounts. These measures can provide additional layers of security beyond password strength, helping to safeguard accounts against unauthorised access and potential breaches.

Ultimately, the reliance on passwords as the main authentication method may decrease as national and global regulations start requiring the use of state-of-the-art technology to protect assets, such as personal data. This shift could lead to the growing adoption of advanced biometric solutions, such as NFC chip implants [9]. However, as our research has shown, users' overconfidence in their own security practices may hinder their willingness to adopt more secure solutions. Addressing this perception gap through education and awareness initiatives will be crucial in fostering acceptance of more robust security measures in the future.

References

1. Abbott, J., Patil, S.: How mandatory second factor affects the authentication user experience. In: Proceedings of the 2020 CHI Conference on Human Factors in Computing Systems (2020). https://doi.org/10.1145/3313831.3376457
2. Abiew, N.A.K., Banning, S.O., et al.: Design and implementation of cost effective multi-factor authentication framework for ATM systems. Asian J. Res. Comput. Sci. **5**(3), 7–20 (2020)
3. Adams, A., Sasse, M.A.: Users are not the enemy. Commun. ACM **42**(12), 40–46 (1999). https://doi.org/10.1145/322796.322806
4. Alodhyani, F., Theodorakopoulos, G., Reinecke, P.: Password managers-it's all about trust and transparency. Future Internet **12**(11) (2020). https://doi.org/10.3390/fi12110189
5. Alroomi, S., Li, F., ACM: measuring website password creation policies at scale. In: 30th ACM SIGSAC Conference on Computer and Communications Security (ACM CCS), Copenhagen, Denmark, pp. 3108–3122 (2023). https://doi.org/10.1145/3576915.3623156
6. Amador, J., et al.: Prospects for improving password selection. In: 19th Symposium on Usable Privacy and Security (SOUPS), Anaheim, CA, pp. 263–282 (2023)

7. Amft, S., Holtervennhoff, S., Huaman, N., Acar, Y., Fahl, S.: Would you give the same priority to the bank and a game? I do not! Exploring credential management strategies and obstacles during password manager setup. In: Proceedings of the Nineteenth Symposium on Usable Privacy and Security, Soups 2023, pp. 171–190 (2023)
8. Amft, S., et al.: We've disabled MFA for you: an evaluation of the security and usability of multi-factor authentication recovery deployments. In: CCS 2023 - Proceedings of the 2023 ACM SIGSAC Conference on Computer and Communications Security, pp. 3138–3152 (2023). https://doi.org/10.1145/3576915.3623180
9. BioTeq Limited: BioTeq are the UK's leading human technology implant specialists (2024). https://www.bioteq.co.uk/
10. Bitwarden: World password day survey 2023 (2023). https://bitwarden.com/resources/world-password-day/#2022-world-password-day-results
11. Bitwarden: World password day – global survey 2024 (2024) https://bitwarden.com/resources/world-password-day/#overview
12. Bonneau, J., Herley, C., van Oorschot, P.C., Stajano, F.: The quest to replace passwords: a framework for comparative evaluation of web authentication schemes. In: 2012 IEEE Symposium on Security and Privacy, pp. 553–567 (2012). https://doi.org/10.1109/SP.2012.44
13. British Standards Institute: BS ISO/IEC 29115:2013: Information technology. Security Techniques, Entity Authentication Assurance Framework (2013)
14. Brockbanks, P., Butler, M.J.: The one-eyed leading the blind: understanding differences between IT professionals and non-IT staff when creating and managing passwords. In: 15th IFIP WG 11.12 International Symposium on Human Aspects of Information Security and Assurance (HAISA). IFIP Advances in Information and Communication Technology, vol. 613, pp. 211–222. Electr Network (2021). https://doi.org/10.1007/978-3-030-81111-2_18
15. Carroll, L.: Alice's Adventures in Wonderland. Sam'l Gabriel Sons & Company, New York, USA (2006)
16. CESG, Centre for the Protection of National Infrastructure: Password guidance simplifying your approach. CESG (The Information Security Arm of GCHQ) and CPNI (Centre for the Protection of National infrastructure) (2015)
17. Christmann, M., Mayer, P., Volkamer, M., ACM: vision: what Johnny learns about password security from videos posted on YouTube. In: Proceedings of the 2021 European Symposium on Usable Security, EUROUSEC 2021, pp. 124–128 (2021). https://doi.org/10.1145/3481357.3481528
18. Clark, J.: ID4D Practitioner's Guide, Technical report, World Bank Group (2019)
19. Dell'Amico, M., Michiardi, P., Roudier, Y.: Password strength: an empirical analysis. In: 2010 Proceedings IEEE INFOCOM, pp. 1–9 (2010). https://doi.org/10.1109/INFCOM.2010.5461951
20. Dumaru, A., et al.: I have my dad, sister, brother, and mom's password': unveiling users' mental models of security and privacy-preserving tools. Inf. Comput. Secur. **32**(3) (2024)
21. Florencio, D., Herley, C.: A large-scale study of web password habits. In: Proceedings of the 16th International Conference on World Wide Web, pp. 657–666. WWW'07, Association for Computing Machinery, New York, NY, USA (2007). https://doi.org/10.1145/1242572.1242661
22. Gautam, A., Lalani, S., Ruoti, S.: Improving password generation through the design of a password composition policy description language. In: Eighteenth Symposium on Usable Privacy and Security (SOUPS 2022), pp. 541–560. USENIX Association, Boston, MA (2022)

23. Gerlitz, E., Häring, M., Smith, M.: Please do not use !? Or your license plate number: analyzing password policies in German companies. In: Seventeenth Symposium on Usable Privacy and Security (SOUPS 2021), pp. 17–36. USENIX Association (2021)
24. Gerlitz, E., Häring, M., Smith, M., Tiefenau, C.: Evolution of password expiry in companies: measuring the adoption of recommendations by the German federal office for information security. In: Nineteenth Symposium on Usable Privacy and Security (SOUPS 2023), pp. 191–210. USENIX Association, Anaheim, CA (2023)
25. Grassi, P., et al.: Digital Identity Guidelines (2017)
26. Gray, C.M., Santos, C.T., Bielova, N., Mildner, T.: An ontology of dark patterns knowledge: foundations, definitions, and a pathway for shared knowledge-building. In: Proceedings of the CHI Conference on Human Factors in Computing Systems. CHI '24, Association for Computing Machinery, New York, NY, USA (2024). https://doi.org/10.1145/3613904.3642436
27. Guo, Y., Zhang, Z., Guo, Y., Guo, X.: Nudging personalized password policies by understanding users' personality. Comput. Secur. **94** (2020). https://doi.org/10.1016/j.cose.2020.101801
28. Henricks, A., Kettani, H.: On data protection using multi-factor authentication. In: Proceedings of the 2019 International Conference on Information System and System Management, pp. 1–4. ISSM 2019, Association for Computing Machinery, New York, NY, USA (2020). https://doi.org/10.1145/3394788.3394789
29. HM Government: Government Cyber Security Strategy, Technical report, Cabinet Office (2021)
30. Humaidi, N., Alghazo, S.H.A.: Procedural information security countermeasure awareness and cybersecurity protection motivation in enhancing employee's cybersecurity protective behaviour. In: 10th International Symposium on Digital Forensics and Security (ISDFS). Maltepe, Turkey (2022). https://doi.org/10.1109/isdfs55398.2022.9800834
31. Kallinikos, J.: The order of technology: complexity and control in a connected world. Inf. Organ. **15**(3), 185–202 (2005). https://doi.org/10.1016/j.infoandorg.2005.02.001
32. Kusyanti, A., Ayu Catherina, H.P., Lia Sari, Y.A.: Protecting Facebook password: Indonesian users' motivation. Procedia Comput. Sci. **161**, 1182–1190 (2019). https://doi.org/10.1016/j.procs.2019.11.231
33. Lee, K., Sjöberg, S., Narayanan, A.: Password policies of most top websites fail to follow best practices. In: Eighteenth Symposium on Usable Privacy and Security (SOUPS 2022), pp. 561–580. USENIX Association, Boston, MA (2022)
34. Mayer, P., Munyendo, C., Mazurek, M., Aviv, A.: Why users (don't) use password managers at a large educational institution. In: Proceedings of the 31ST USENIX Security Symposium, pp. 1849–1866 (2022)
35. Mazurek, M.L., et al.: Measuring password guessability for an entire university. In: Proceedings of the 2013 ACM SIGSAC Conference on Computer and Communications Security, pp. 173–186. CCS '13, Association for Computing Machinery, New York, NY, USA (2013). https://doi.org/10.1145/2508859.2516726
36. Mennes, F.: 2024: the age of a workforce free from passwords (2024). https://www.techradar.com/pro/2024-the-age-of-a-workforce-free-from-passwords
37. Micallef, N., Arachchilage, N.A.G.: Changing users' security behaviour towards security questions: a game based learning approach. In: 2017 Military Communications and Information Systems Conference (MilCIS), pp. 1–6 (2017). https://doi.org/10.1109/MilCIS.2017.8190424

38. Morris, R., Thompson, K.: Password security: a case history. Commun. ACM **22**(11), 594–597 (1979). https://doi.org/10.1145/359168.359172
39. Mwagwabi, F., Jiow, J.H.: Compliance with security guidelines in teenagers: the conflicting role of peer influence and personal norms. Australas. J. Inf. Syst. **25** (2021). https://doi.org/10.3127/ajis.v25i0.2953
40. National Cyber Security Centre: NCSC lifts lid on three random words password logic (2021). https://www.ncsc.gov.uk/news/ncsc-lifts-lid-on-three-random-words-password-logic
41. National Cyber Security Centre: Cyber essentials: requirements for IT infrastructure (2022)
42. Olade, I., Liang, H., Fleming, C.: Story-based authentication for mobile devices using semantically-linked images. Int. J. Hum. Comput. Stud. **171** (2023). https://doi.org/10.1016/j.ijhcs.2022.102967
43. Peer, E., Egelman, S., Harbach, M., Malkin, N., Mathur, A., Frik, A.: Nudge me right: personalizing online security nudges to people's decision-making styles. Comput. Hum. Behav. **109** (2020). https://doi.org/10.1016/j.chb.2020.106347
44. Qu, L., Xiao, R., Shi, W., Huang, K., Qin, B., Liang, B.: Your behaviors reveal what you need: a practical scheme based on user behaviors for personalized security nudges. Comput. Secur. **122**, 102891 (2022). https://doi.org/10.1016/j.cose.2022.102891
45. Ray, P., Giri, D., Meng, W., Hore, S.: GPOD: an efficient and secure graphical password authentication system by fast object detection. Multimedia Tools Appl. (2023). https://doi.org/10.1007/s11042-023-17571-4
46. Rogers, R.W.: Cognitive and physiological processes in fear appeals and attitude change: a revised theory of protection motivation. In: Social Psychology: A Source Book, pp. 153–176 (1983)
47. Shay, R., et al.: Correct horse battery staple: exploring the usability of system-assigned passphrases. In: Proceedings of the Eighth Symposium on Usable Privacy and Security. SOUPS '12, Association for Computing Machinery, New York, NY, USA (2012). https://doi.org/10.1145/2335356.2335366
48. Stylios, I., Kokolakis, S., Thanou, O., Chatzis, S.: Key factors driving the adoption of behavioral biometrics and continuous authentication technology: an empirical research. Inf. Comput. Secur. **30**(4), 562–582 (2022). https://doi.org/10.1108/ics-08-2021-0124
49. Thaler, R.H., Sunstein, C.R.: Nudge : Improving Decisions about Health, Wealth and Happiness. Penguin Books, London (2009)
50. Tsoukas, V., Gkogkidis, A., Kakarountas, A.: A survey on mobile user perceptions of sensitive data and authentication methods. In: Proceedings of the 24th Pan-Hellenic Conference on Informatics, pp. 346–349 (2020). https://doi.org/10.1145/3437120.3437337
51. Wang, V., Tucker, J.: Surveillance and identity: conceptual framework and formal models. J. Cybersecurity **3**(3), 145–158 (2017)
52. Wang, V., Tucker, J.: I am not a number: conceptualising identity in digital surveillance. Technol. Soc. **67** (2021)
53. Woods, N., Siponen, M.: How memory anxiety can influence password security behavior. Comput. Secur. **137**, 1–24 (2024). https://doi.org/10.1016/j.cose.2023.103589
54. Wu, X., Munyendo, C., Cosic, E., Flynn, G., Legault, O., Aviv, A.: Assoc computing machinery: user perceptions of five-word passwords. In: Proceedings of the 38th Annual Computer Security Applications Conference, ACSAC 2022, pp. 605–618 (2022). https://doi.org/10.1145/3564625.3567981

55. Xie, Z., et al.: GuessFuse: hybrid password guessing with multi-view. IEEE Trans. Inf. Forensics Secur. **19**, 4215–4230 (2024). https://doi.org/10.1109/TIFS.2024.3376246
56. Zhou, T., Wu, H., Lu, H., Xu, P., Cheung, Y.: Password guessing based on GAN with Gumbel-Softmax. Secur. Commun. Netw. **2022** (2022). https://doi.org/10.1155/2022/5670629
57. Zimmermann, V., Marky, K., Renaud, K.: Hybrid password meters for more secure passwords - a comprehensive study of password meters including nudges and password information. Behav. Inf. Technol. **42**(6), 700–743 (2023). https://doi.org/10.1080/0144929x.2022.2042384

Principles for Inclusive Assessment Design in Cybersecurity Education

Rosanne English[1(✉)], Joseph Maguire[2], and Adriana Wilde[3]

[1] University of Strathclyde, Glasgow, UK
rosanne.english@strath.ac.uk
[2] University of Glasgow, Glasgow, UK
joseph.maguire@glasgow.ac.uk
[3] University of Southampton, Southampton, UK
a.wilde@soton.ac.uk

Abstract. When designing cybersecurity assessments, educators aim to achieve several objectives. Primarily, we need to measure how well students meet the intended learning outcomes while ensuring that the assessments are both authentic and valid. This involves providing opportunities for students to develop their security skills and knowledge. Achieving these objectives is challenging and often requires substantial time and effort. In addition to these goals, we must also ensure that the assessments are inclusive since, arguably, a non-inclusive assessment is not truly valid. In cybersecurity, designing inclusive assessments without inadvertently omitting key knowledge and skills can be particularly difficult. This chapter reflects on relevant research on inclusive assessment and identifies core principles that can be applied to cybersecurity assessment design in higher education to improve inclusivity.

1 Introduction

In cybersecurity education, careful attention must be given to both assessment design and the creation of feedback opportunities to support students in achieving the intended learning outcomes. For students, assessment serves as a motivator to engage with the material, helping them to discern the key aspects of a topic and gauge their understanding. Feedback is essential for students to understand their next steps following an assessment. It helps them identify gaps in their knowledge and skills, guiding their subsequent efforts.

For educators, assessment acts as a tool to ensure essential material is positioned to support the learner's desired knowledge and skill development. It also provides insight into how learners have interpreted and used material and resources. Feedback provides an opportunity for the educator to communicate directly to the learner their strengths and areas of improvement with respect to the learning objectives. Assessment, by its nature, is exclusionary in some sense: it excludes learners from progressing if they have not demonstrated the intended learning outcomes of a given course. However, educators must ensure

P. Legg et al. (Eds.): CSE 2024, LNNS 1213, pp. 192–212, 2024.
https://doi.org/10.1007/978-3-031-77524-6_11

assessment is not unnecessarily exclusionary. Deterring learners due to poor performance on irrelevant factors for a given topic only serves to drive learners away from a topic. This is evidenced in advice around what constitutes 'good' feedback, such as Nicol and Macfarlane-Dick seven principles which highlight that feedback should be in line with the predefined criteria of assessment [29].

Inclusive assessment is typically thought of as ensuring there is equitable opportunity for all learners to demonstrate achievement of learning outcomes (as discussed by Christine Hockings and Terentjevs [8]). In simpler terms, inclusive assessment is the removal of unnecessary barriers to demonstrating achievement. An unnecessary barrier can be perceived, at best, as an inefficiency in the assessment design and, at worst, as a contaminant that compromises the validity of the assessment itself. Consequently, removing those unnecessary barriers through careful assessment design should not only (1) produce a more efficient and effective assessment design, it should also (2) ensure the assessment is itself more inclusive, supporting and nurturing a more diverse culture of learners. The challenge for assessment design is to ensure unnecessary barriers are removed whilst ensuring the assessment remains effective. Further, and specifically within the British Higher Education context, accrediting bodies are increasingly requiring that departments in engineering disciplines promote "embedding inclusive design within the curriculum where relevant" [10, 45].

Whilst much of the above applies to many disciplines, cybersecurity education presents additional challenges [47], emerging from the speed of change and innovation in such a dynamic landscape. In this context, an emerging concern is that educators, in an effort to ensure inclusivity, may consequently favour *general principles* of inclusive design rather than reflecting what is critical or core to cybersecurity and what must be maintained in the assessment design.

In this chapter, we will explore key principles of inclusive assessment in cybersecurity by reflecting on forms of assessment, what inclusive assessment means in higher education, and identify the key principles to ensure inclusive assessment design in a cybersecurity context. We will conclude by introducing an example assessment and reflecting on the next steps for the cyber education community.

2 Assessment in Higher Education

For educators, an assessment may be considered an instrument or support structure for learners that supports them in their progress. For the learner, an assessment should be a valuable indicator to them of the gap that exists in terms of their current knowledge or skill level and their intended final goal. The first aim of assessment or primary outcome should be to communicate to the learner the progress they have made on their learning journey and what progress is still to be made. An assessment should provide sufficient insight to the learner for them to appreciate the *gap* that exists between what they have attained and their final goal. The secondary aim or outcome of assessment is to communicate to everyone else, the same information. The difference is what an observer, other than the learner does with this information.

From this perspective of learning, assessment could be considered as assessment *for* learning [42] rather than *of* learning. For other observers, such as employers and educators, the outcome of assessment may be used for alternative uses. For an employer, assessment outcome may be just be another outcome to select a potential applicant or in terms of selecting an appropriate role or task. For the educator, assessment outcomes could be used to support the individual or potentially interpret the aggregate of outcomes from a cohort of learners to revise a course.

Consider, for example, a lecturer or educator in an introductory cryptography course who has created a series of assessments contributing to the final grade. If the overall cohort performance is medium to high, with most students achieving high grades, this might be interpreted as a positive outcome, indicating that the course is well-structured and aligned, effectively meeting its aims and objectives. Students would be equipped with a strong foundational knowledge of cryptography, providing a solid basis for subsequent courses.

Alternatively, the educator might conclude that students are not being sufficiently challenged and that a more varied performance distribution is desirable. This could lead to expanding the course's scope, increasing the level of difficulty, and observing the impact on the next cohort.

Another possibility is for the educator to perceive the course as successful and maintain its structure for several years. However, when students reach their third or fourth year, other educators might question the depth of their foundational cryptography knowledge. This discrepancy could emerge from assessments designed under the assumption of strong foundational understanding. Poor performance in these later assessments might not reflect the quality of the assessment itself but rather the inadequacies of earlier assessments, the effects of which become apparent only after several years.

The current illustration presupposes the course coordinator and colleagues have faith in the assessments within the course themselves. The assessments are working effectively and producing strong cohort performance. However, if we consider from the principle that the broad aim of assessment is to measure *something*, then at a minimum, all individuals involved in the assessment must have confidence in the instrument that is used to measure.

The measurement instrument or assessment must be *reliable* and *valid*. The assessment must be reliable in that outcomes are similar across repeated use and are valid. Valid in that the measurement is an appropriate indicator of what is being measured. The tenants of reliability and validity are core to assessment, and it is valuable to consider both prior to reflecting on whether an assessment should be labelled as inclusive or not.

For many assessments in higher education, confidence can be reasonably strong and well placed for how effective they are in measuring 'something'. Assessments that test largely declarative and procedural knowledge are expected to be predictable and reliable. It is reasonable to assume that the first few semesters or the foundational years within a programme can be covered by such an assessment. Assessment where educators and learners can have a high degree of confidence.

Consequently, reflecting on our previous example, an assessment in a fundamental cryptography course should be fairly reliable in producing reliable outcomes. However, as learners progress further into higher education, the reliability and validity of assessment become less dependable for a given context. Carmines and Zeller emphasise that an effective assessment is an effective measurement in that it relates to some skill, knowledge or concept back to an empirical indicator [7].

Again, reflecting on our previous example, learning to define an effective assessment, a measurement instrument that relates back to a more elaborate concept or application of knowledge within an advanced course, could be more challenging in that assessment reliably produces the same outcomes for every execution.

If we accept that an effective assessment is an effective measurement of some skill, knowledge, or essentially a concept, it is related back to an empirical indicator [7]. From this perspective, when considering something as an empirical measurement, we must have confidence that the measurement is reliable and valid.

We should consider reliability first before validity. Reliability can be considered as the accuracy of the instrument used to measure, in terms of output after each attempt. It is reasonable to conclude that a measurement instrument, from our perspective for an assessment, could be considered reliable if the same result or a similar result is obtained upon subsequent attempts.

Carmines and Zeller emphasises that when measuring a phenomenon no instrument will always be predictable, there will be an element of unpredictability [7]. The expectation is the same exact result may not be obtainable but that there will be relative consistency.

Carmines and Zeller provide the example of blood pressure; if we measure the blood pressure of a cohort of individuals on a particular day, the measurements may not be exactly the same the following day. However, the individual with the highest blood pressure on a given day would be expected to be near the top of the list the following day and the day after that. The expectation is that while the exact measurement may fluctuate, the consistency would remain relative, and the expectation is the individual would remain near the top of the list even as the cohort increases [7].

Therefore, an effective measurement, that is, an effective assessment, must produce reliable and predictable outcomes. While those outcomes may not be exact, there is acceptance that there will be some element of unpredictability, but that is within acceptable bounds and consistent.

An assessor must also have confidence in its validity to be confident in its reliability. There is little rationale or motivation for having an assessment that reliably produces outcomes, but the outcome is not an effective measurement of what is being assessed.

Educators may consider an examination an effective assessment of a learner's knowledge and skills. However, Sambell et al. reports that some students perceive exams as a test of memory, stamina, stress and luck rather than the knowledge and skills [38]. Consequently, even an assessment that can be demonstrated to assess or measure something specific may not be perceived as such by those engaging with the assessment.

Let us reflect on our example of the assessment in a fundamental cryptography course. The assessment may be delivering consistent, reliable results, but what it measures is not a good indicator of fundamental knowledge of cryptography. It could just be a good indicator of coping with stress, time management and/or memory. It may be a strong performance on the assessment, while consistent and reliable, so the top performer is likely to be among the top performers the following day and the next, but the outcome is not valid. Valid in the sense that it is not a strong indicator of knowledge of fundamental cryptography.

Therefore, validity and reliability are additional core aspects of assessment. Validity can be considered as the effectiveness of an assessment in terms of measuring what it intends to measure. Validity is not binary; the validity of the assessment is not true or false but in terms of how effective an assessment is at relating back to an indicator or measurement [12].

Carmines and Zeller highlights the example of a driving test, a valid assessment may contain a practical and theory element whereas an assessment that lacks either may not be particularly effective or valid in terms of relating an individual back to the indicator of pass or fail [7].

Downing argues that validity is more a *construct validity*, that is the assessment relates to underlying theoretical concepts. As such, the assessment should be approached as a hypothesis with multiple sources of evidence provided to accept or reject that hypothesis [12].

Consequently, the more 'high-stakes' the assessment, such as an 80% exam, the more evidence is required to confirm that the assessment is valid. Valid in terms of how performance on the assessment is being interpreted and related back to the course aims and outcomes. If a given score on a test is being interpreted as a pass or fail, then sufficient evidence needs to be provided to accept that interpretation. Messick argues that evidence of validity is always incomplete, and as such, the argument being made is always the "best case" that can be made at a particular point in time [26].

If we relate this back to our example, while an assessment in a fundamental cryptography course may be reliable and not valid from the perspective of the educator of senior students, it is also possible, potentially more likely, that the assessment in the senior years is not reliable and not valid. A position that could be possible in that *reliable* assessing, say, more theoretical and conceptual material is more challenging in terms of relating it back to empirical measurement, but also difficult to argue that it is valid unless there a reasonable preponderance of evidence to suggest as such.

Consequently, all educators need to reflect on whether an assessment is reliable and valid prior to every execution to some extent. However, the concepts of reliability and validity also infer whether an assessment is inclusive or not inclusive. If we take our example, the assessment in the senior year cryptography could be considered a non-inclusive assessment.

The course coordinator has implicitly assumed, not unjustifiably, that learners have a strong fundamental grasp of cryptography. This assumption is due to performance on previous assessments. However, the reality is that students do not have a fundamental grasp of cryptography, and so perform poorly on the assessment. The assessment could be reliable and valid if the initial assumption holds that a student with a strong knowledge of cryptography performs well in the assessment. However, if they do not, the student will perform poorly on the topic.

The student may interpret the outcome as not having sufficient understanding of the senior topic when the actual lesson is that they do not have a fundamental grasp of cryptography. This assessment could be labelled as non-inclusive, and for the reader, this could be a challenging perspective to accept.

An alternative example could be if the educator of the senior security course decided they wanted students to write an essay to assess their knowledge of a given topic. The educator could make the assumption that learners in a senior security course have a strong written ability and a strong grasp of the language. However, this may not be the case for all students; again, poor performance on the assessment may not be due to poor writing and communication skills, not because they have not learned what is expected in the course aims and objectives.

From this perspective, such a written assessment could be deemed to be a *non-inclusive* assessment. If we consider the assessment to be reliable and valid, it presents a potentially unnecessary barrier in that those with poor written and communication skills will perform poorly.

Consequently, an educator may take different routes within the assessment to address this issue. Alternatively, an educator may deem written communication part of the assessment, and students should be more than proficient in these skills. This is a reasonable position, and the educator may want to be transparent or clear about it, partly to support the learner in understanding what the outcome of the assessment communicates. This relates to the inclusive principle of transparency.

Therefore, an assessment can be reliable and valid but deemed to be non-inclusive. To strengthen that assessment even further, educators will want to reflect on the implicit and/or explicit assumptions they make in its design. In the next section, we consider inclusive assessment in more detail and corresponding actions that can be taken to reduce non-inclusive assessment.

3 Inclusive Assessment in Higher Education

Reliability and validity are central to the effectiveness of assessment. However, an assessment could be considered reliable and valid, but a learner could demonstrate poor performance that is more indicative of issues elsewhere. For example, an educator could set an assessment to write a critical essay on the topic of fileless malware and make the assumption that for a given level, a learner would have strong written communication skills.

If this is not the case, the learner may exhibit poor performance as they are poor at writing critical essays. Such a performance does not mean that the learner knows little about the given topic or how to critique it. The educator and learner may interpret the outcome of the assessment in the same way, that is the learner knows little about the topic or what they do know, is wrong.

The problem is that the need to write a critical essay is potentially a barrier to demonstrating the skill and knowledge the educator is looking for from the learner. In this way, the assessment is not inclusive, it presents an unnecessary barrier. This will not be true in every instance, but it is worthwhile for the educator to reflect and decide what potential barriers may exist in demonstrating strong performance in assessment, but are not necessarily required.

Rose highlights a number of important aspects around the *accuracy* of assessment if we consider the aim is to assess learning, outlining areas for consideration such as limited means of expression, representation, and engagement [37].

In terms of *limited means of expression*, the concern is that assessment generally is structured or constructed in such a way that it is one vehicle, that there is only one option. If we then consider the application of that method or vehicle, it could be applied to an entire class or cohort.

The approach could be an oral examination or a written assessment, an assessment could also be constructed as a skills assessment. The concern is that for individuals who have difficulties hearing or language difficulties, where the language of the assessment is not their first language, then the assessment is arguably not only assessing the topic but also the ability to engage in an oral examination. The response could be to offer a written approach, but the same difficulty will apply; that is, there will potentially be at least one student who has difficulty in the assessment media. From the outset, a useful aspect for an educator to retain is that the accuracy of an assessment will be undermined if there is only one way to participate. This is only likely to increase as the cohort becomes bigger and bigger.

For *limited means of representation*, there is also the challenge around representation. An individual may not be able to represent themselves sufficiently to perform at their best at an assessment. This could scale from simply not being able to physically write or speak an answer to being faster or slower at keyboard entry.

Again, from the perspective of education, it is worthwhile to consider how much the approach to expression impacts the accuracy of the assessment. If a student cannot write fast enough or not at all, or cannot type fast enough, how much of the assessment reflects this limited means of expression rather than knowledge or skills of a given topic? If the only means of expression is the same for the whole cohort, then it is likely the assessment will not be accurate. This will only become further undermined as the class scales.

If we can consider *limited means of engagement*, high and low engagement with an assessment are unlikely to produce accurate assessment results. Reflecting on the example of a 2-hour exam paper in a 24-hour window, a student engaging with an assessment for 15 min is unlikely to produce positive results and a student engaging with the assessment, non-stop, for 24 hours is similarly unlikely to produce results better than, say, 3–4 hours.

The accepted wisdom is that (a) some level of engagement is necessary with assessment and (b) there is an optimal level of engagement. The concern is that the rewards for the assessment are typically the same for all students, as the reward has to be sufficient to motivate most learners to engage.

However, such significant rewards are likely to produce a reaction, say anxiety, that will impact the individual's ability to perform well and undermine the accuracy of the assessment. Consequently, again, having the same rewards for all students is likely to have or lead to undesirable impacts that undermine the accuracy of the assessment. The solution is not obvious, but recognising such a limitation is to recognise the limitation in the accuracy of the assessment on the skills and knowledge of the individual.

We can see from these areas that there is a thread of inclusivity, though it is not explicitly addressed. When considering the literature on inclusive assessment, one approach which is put forward as a foundation or possible solution for inclusive assessment is that of Universal Design for Learning proposed by Orkwis and McLane [31]. This is a framework in which the goal is to design learning which is accessible for all learners without the need for further adjustment. Core principles include:

1. Multiple means of engagement i.e. designing modules to motivate and suitably challenge learners
2. Multiple means of representation i.e. presenting information using multiple modalities
3. Multiple means of action and expression, e.g. different ways for learners to express their learning, providing multiple routes for student action such as different forms of communication

However, whilst there is increasing conversation about the use of inclusive assessment design to compensate for or address a number of problems in higher education, there is little reported evidence on the impact in terms of student learning [41].

Norwich argues that adjusting and adapting assessments for different students or groups of students to support their requirements is not inclusive design [30]. Instead, Norwich argues that inclusive design of assessment includes *all* students and that educators face a dilemma in delivering effective assessment that meets the needs of all learners without having unnecessary barriers. This is arguably in agreement with the concept of universal design and aligns with the idea that non-inclusive assessment is, by its nature, an ineffective assessment.

However, inclusive assessment can be difficult to define. Bain et al. reports that inclusive assessment can be considered an assessment that enables learners to achieve outcomes without adapted or alternative assessments [4]. We will use this as our working definition. The literature reports many different approaches and principles to inclusive assessment. However, some common themes are choice, transparency, and flexibility. We now consider each of these in turn.

3.1 Choice

Student choice is one often promoted approach for inclusive assessment. In this approach, students are afforded control or options around some aspect of the assessment [28]. The level of choice can vary from being considered relatively small, for example, selecting an essay or presentation topic from a predetermined list, to being significant and large, such as alternative routes through an assessment, for example, either producing the output as a written report or programming exercise [23]. O'Neill argues that providing choice in assessments is challenging as many learners and educators may perceive the choice as unfair [32]. Having said that, O'Neill suggests that limited choice does have a positive impact on learners when compared to learners that performed the same assessment but without limited choice [32].

3.2 Transparency

Ensuring students comprehend and understand what is being assessed is deemed another characteristic of inclusive assessment. Many educators would argue that it is obvious what is being assessed but it could be due to prior knowledge and experience of the learner, they do not have a clear appreciation of what is being assessed. Parker argues that learners must be aware of what will be assessed by educators and how the process will happen [33]. It is also a valuable reflection point for educators to consider what their expectations of the learner are beyond their classroom. In many ways, the issue of transparency could be framed as assessment literacy.

Assessment literacy can be considered as the process of comprehending and understanding academic judgement [9]. The expectation is that an individual performing an assessment has internal expectations or at least an idea of the standards and criteria that are acceptable [35]. If an individual was completing an essay in the first year of their programme, they may make a mistake as they have poor assessment literacy of an essay at university-level as prior experience may be that of submitting essays at school. The expectation is that by the time an individual reaches the final year of their programme, they would have stronger literacy or assessment literacy around the expectations of an essay [16].

An individual who performs poorly in the final year due to poor assessment literacy is likely to feel frustrated and disappointed as they have not fully appreciated how they will be assessed [25]. In many ways, this can be addressed by providing sufficient opportunities to fail as part of their programme of learning [24]. If an individual is given many opportunities to write essays in their early years, they can develop an understanding of and appreciation of the expectations of that assessment.

The same can be considered of inclusive and transparent assessment. An educator may have set an assessment, such as an essay, that they would have deemed as an inclusive assessment. Partly in the expectation that a senior student would have sufficient experience in writing such essays, i.e. they have strong assessment literacy for essays. However, if we have a student who does not have such experience or assessment literacy, then the assessment would not be considered inclusive. In that it has presented unnecessary barriers, the educator has made assumptions in the design of the assessment that the learner is sufficiently aware of the standards of written work. This leads to another aspect of inclusive assessment, that is programmatic assessment.

3.3 Programmatic Assessment

Programmatic assessment can be considered a programme of assessment where a collection of assessments contribute to the assessment of the learner rather than a few high-stakes assessments [43]. Knight argues that assessments need to be reliable and valid, but such qualities often result in such assessments being coarse, simplistic and high-stakes [19]. The concern is that such high-stake assessments, by their nature, are not inclusive as they are devised to ensure they remove inference and judgement against criteria and are instead approach a binary choice that is more transferable [19].

Instead, an approach can be adopted where the culmination of assessments can be used to assess a learner through several assessments rather than relying on a single or few high-stakes assessments.

Schuwirth et al. suggests that such an approach is becoming more common in medical education across the world [39]. The aim is that rather than assessing being *of* learning, it is for the purposes of learning in of itself [39]. Schuwirth et al. outlines a plan for learning where an individual is provided with a series of assessments over several courses, with the focus being the delivery of meaningful feedback to the learner for each of those assessments.

The learner would then be required to collect the results of the assessments, feedback, and reflections in a portfolio. The portfolio would then be assessed by an educator or a panel of educators to determine the final over-arching grade for the programme. Consequently, rather than the learner completing a single or few high-stakes assessments at the end of a programme, they would complete several assessments over the course of the programme. The panel of educators could then consider the collection of these assessments to make an assessment of the individual.

Such a process could be considered more inclusive, as the learner is provided multiple opportunities in different assessments to drive their learning rather than focusing on a few assessments that focus on demonstrating learning.

3.4 Flexibility

Irwin and Hepplestone argues that flexibility in assessment can take many forms; while it can be broadly considered the same as choice, it could also take other forms in assessment [17]. In particular, Irwin and Hepplestone highlights a particular case study where students were able to select the grades from assessments that contribute to their final award [40]. In many ways, giving students the flexibility to select their grades from components would suggest they would select those components where they perform well. Consequently, the anxiety where the reward is the same for all assessments is reduced as the reward is not the same.

Flexibility around assessment is an approach that could support assessment being more inclusive. Flexibility could also apply to where the assessment happens; the assessment could be completed on-campus, or it could be completed remotely from another location, which may reduce stress for learners.

Having identified a working definition of inclusive assessment and relevant principles that can be considered in designing it, we will now consider assessment design in a cybersecurity context in the following section.

4 Assessment in Cybersecurity

Köhler and Meinel outlines the range of assessment types from the perspective of establishing the understanding of employees. This includes quizzes, questionnaires, interviews, observations, monitoring of activity and technical exercises such as red teams [21].

Many of these aspects are reflected in higher education. Typical formats of assessment in higher education include coursework, projects, reports, presentations, quizzes, and unseen exams.

Often in higher education, the standard mechanism for assessment is a written exam. This is a familiar and well-established assessment tool. As outlined by French et al., there is evidence that high-stakes unseen exams can have the benefits of increasing memory and recall, increasing student motivation, providing a benchmark across a cohort, and increasing confidence in the authorship of the candidates' work [14]. However, there are also common critiques of this approach. Exams knowledge retention is often short-lived, can focus on simple knowledge recall and is argued, in most areas, non-authentic [14]. This is particularly valid in a cybersecurity context, where exam assessment is not generally authentic (with the exception of providing experience for completing professional certification exams such as CISSP). Exams can also potentially present a barrier to students who may (given an alternative format) be better able to demonstrate learning outcomes due to issues such as anxiety [14].

In contrast, and as illustrated in Table 1, coursework offers greater opportunities for authentic experiences and can facilitate hands-on security exposure in areas such as penetration testing, defence exercises, and secure programming. A drawback is the limited capacity of learners to undertake such assessments across multiple modules within a short time frame (typically a ten to twelve-week semester), particularly when deadlines tend to cluster. Furthermore, with increasing student numbers, effective assessment by educators becomes more challenging, potentially diminishing the impact of valuable feedback due to delays. There is also a concern that learners' true comprehension might not be accurately assessed given the allotted time. This challenge is exacerbated by recent advancements in technologies such as generative AI.

One alternative approach is to prioritize the assessment of skills over outputs, which can be closely aligned with higher-order cognitive skills as outlined

Table 1. Comparison between exam and coursework as assessment methods

Type	By whom?	Examples	Benefits	Drawbacks
Exam	Individual	hand-written tests, computer-aided assessments, multiple-choice questions, oral exams.	Familiar to students and staff, often easy to scale, reusable, controlled conditions, ideal to test knowledge.	Lacks authenticity, highly-pressured. not ideal to test skills.
Coursework	Groups or individual	Capture The Flag (CTF), portfolios, coding projects, digital forensics investigations, penetration testing challenges, reports,	Authentic, video presentations. practical/hands-on, easier to link with employability skills.	Harder to prove authorship. Harder to scale, susceptible to abuse of GenAI, Limited re-usability.

in Bloom's taxonomy [5,6]. Skills are frequently highly valued by employers and can be well-suited to a competency-based framework. However, these can be challenging to assess comprehensively and efficiently, and certain skills may exhibit limited "shelf life" beyond a five-year time frame (and perhaps an even smaller time frame for cyber-specific skills), which contrasts with the extended professional lifespan of graduates.

5 Principles for Inclusive Cyber Assessment

Cybersecurity is undoubtedly a multi-faceted subject encompassing areas such as cryptography, law and regulations, privacy, forensics and many more as reflected in the cybersecurity Body of Knowledge (CYBOK, [1]). LeClair et al. argue that a holistic approach to cybersecurity education is necessary. Consequently, a holistic approach to assessment in cybersecurity education is important [22]. In this section, we will consider the elements of cyber assessments whilst remaining cognisant of inclusive design. We do so by considering the components of focus, outcomes, and the resulting artefact.

5.1 Focus: People, Process or Technology

People, process and technology are often considered the three pillars of cybersecurity [3]. As such, we will use this to reflect on the perspective of the module which should be considered when designing the assessment.

Topics within cybersecurity can be considered from different perspectives. For instance, cryptography can be considered from both a process and technology perspective. Teaching a process perspective of cryptography can include understanding the mathematics behind the RSA algorithm or other protocols. From a technical perspective, it could be asking students to implement encryption within a system using RSA.

It is important to note that both perspectives can be interwoven, one must generally appreciate the process on some level to be able to enact this in practice, but one can understand some elements of cyber from a purely theoretical standpoint. In the example provided, a student would still need to understand the process as they need to know two keys are generated, the requirements of distribution of the keys, as well as padding requirements and context for use in order to apply it appropriately.

In reviewing CyBOK, one can reflect on the different perspectives for the topics (see Table 2).

Table 2. Perspectives of topics in cybersecurity: People, process, technology

CyBOK Topic	Perspectives
Human Organisational	People, Process
Attacks and Defence	Process, Technology
Systems	Process, Technology
Software and Platform	Technology
Infrastructure	Technology

Considering the focus(es) of the module is critical to identifying the essential elements of the assessment which should not be removed when considering other aspects such as inclusive design. For instance, if the focus is on technology such as through an ethical hacking exercise, then one would not adjust the assessment to remove the technical steps involved in an effort to ensure it is accessible.

5.2 Outcomes

Additionally, one must consider the overall objectives of the module - whilst specialist modules exist, cyber modules in undergraduate programmes are often designed to provide a base understanding for students who are not focused on a cyber career [36]. This can inform what type of assessment is completed.

Using a constructive alignment approach [5] one considers the expected outcomes of the module, and then considers which outcomes are to be measured by a given assessment. In line with Bloom's Taxonomy, assessment of security at lower levels is likely to focus on understanding of topics, whilst higher levels are more likely to focus on skills such as judgement and critique [6].

Dawson and Thomson argue that we must look beyond the technical when developing learners' cyber skills [11], emphasising the need to consider people and process. Graham and Lu reviewed 17,929 job adverts in cybersecurity to establish common skills required of these roles and found that skills most sought after were often focused on the 'soft' side – once more emphasising the need to consider people and processes in our assessment design.

Additionally, different contexts (such as the locations) can impact the level of assessment or expectations for a security module. For instance, different countries have different frameworks to reflect levels of study. e.g. the Scottish Credit and Qualifications Framework, the Credit and Qualifications Framework for Wales, the Regulated Qualifications Framework, the US Qualifications Framework or the East African Qualifications Framework for Higher Education to name a few.

Whilst there are differences in nomenclature, often the general expectations are similar. Across the frameworks, we can see generally similar expectations to that of Bloom's revised taxonomy [20] in that as one progresses in qualifications they move up the pyramid, building on skills from lower levels.

Knowing what assessment level should be measured for a given framework can be useful in identifying the types of tasks which one might expect learners to demonstrate for the assessment. For instance, we may have an assessment aiming to evaluate a student's understanding of cyber risk assessment. At a 1st year undergraduate level, this might be to understand the stages in a risk assessment, whilst in the final year it could be to complete an appropriate risk assessment for a given scenario. These are two very different assessments.

Alongside these cyber-specific skills, one should also consider attributes or skills of individuals which are beneficial particularly to cyber e.g. [11] note that the future cyber workforce should be systemic thinkers, team players, have a sense of duty, reflective practitioners, or strong communicators.

If such a skill is considered as fundamental to the assessment, it should be explicitly incorporated into the assessment criteria. For instance, if an assessment is looking to evaluate how well a learner can analyse a given security scenario and communicate this to others, then there should be a defined part of the marking scheme assigned to the quality of communication itself.

5.3 Demonstrating Outcomes: The Artefact

Having considered the focus to identify the objectives and the intended outcomes, such as skills, we now consider how students can demonstrate the outcome. Sometimes, there are limited ways in which one can demonstrate a given outcome, e.g. proof students can successfully attack a vulnerable system can only be fully demonstrated by them doing so, not by a presentation on typical attacks. However, for other areas, there may be alternative outputs which could equally demonstrate the achievement of a given ILO with the defined focus e.g. being able to critique a policy could be demonstrated in an exam, through coursework or by oral presentation of the outcomes.

This can be an area where unnecessary barriers are unintentionally introduced. Whilst a live presentation can assess this well, if a primary output is not presentation skills then this could be setting students with anxiety disorders at a disadvantage. Learners with such challenges may be able to more easily demonstrate the same outcome in a different format. Consequently, one must carefully consider if the format of the artefact is necessary to demonstrate the intended outcomes. For instance, is an exam the only artefact that demonstrates that a student can apply threat modelling to a given scenario? If not, then one may consider alternatives.

In cybersecurity, artefacts can take many forms, including written reports (as discussed in Allison [2]), lab exercises (as seen in Du [13]) capture the flag exercises (as seen in Mirkovic and Peterson [27] and Vykopal et al. [44]), student-produced videos (as in the case study on video coursework for usable security discussed in Wilde et al. [46]), and several other artefacts produced through coursework as listed in Table 1.

5.4 Inclusive Cyber Assessment

Having identified the key elements of cybersecurity assessments:

- Focus
- Intended outcomes
- Artefact

we now consider how inclusive principles can impact design at each stage. From the review in Sect. 3, we identified the following principles which can aid in developing inclusive assessment design:

1. Choice: the ability of learners to have an element of control or choice around some aspect of the choice, e.g. choice of type of artefact used to demonstrate the outcomes
2. Transparency: what is being assessed should be clear to students and other stakeholders (e.g. external examiners)
3. Flexibility: expanding on choice, flexibility in assessment can incorporate greater adjustments such as permitting learners to identify which assessments contribute to their final grade, or flexibility of location to complete the assessment.

We postulate that one can incorporate principles of choice, transparency, and flexibility at each stage when considering principles of choice. We will now consider how these principles may impact each core consideration in cybersecurity.

5.4.1 Focus

Choice. Considering the types of technology-focused assessments in a cybersecurity module, one may come up against accessibility issues for given tools or technology. For instance, some software is not easily adaptable for those who use screen readers. Consequently, we encourage educators to *consider the impact of using software which may be inaccessible for screen readers etc., by considering the choice of technologies used.*

Transparency. Assessments that focus on the 'people' and 'process' elements of cybersecurity are likely to involve analysis, reflection, and communication. Consequently, those with information processing differences, such as Autism Spectrum Disorder (ASD) and Attention-Deficit Hyperactivity Disorder (ADHD), may face additional barriers in completing such assessments.

When designing such assessments, we encourage educators to *consider how to assist students in managing additional challenges*, e.g. incorporating explicit checkpoints into an assessment to aid in time management and task breakdown, for team exercises, consider "a 'team agreement' " document which outlines expected processes where the team disagree and clarifies expectations.

Flexibility. Other technology-focused assessments could include elements of coding. However, the aim may not be to evaluate the quality of the code itself. If the coding element is a means to demonstrate understanding rather than the focus, we encourage educators to *permit use of code generation tools such as copilot* though we would also encourage requiring students to submit the prompts used to generate the code.

5.4.2 Outcomes

Choice. It may be possible to incorporate some choice into the design of the outcomes. This could include elements such as co-designing modules with the student cohort to identify which ILOs should be addressed by a given assessment, though this has some risks and challenges in terms of timelines, quality assurance, and assessment management.

A less cumbersome approach to choice for outcomes is to provide a choice in topic, such that learners can demonstrate a given outcome but choose an area or case study of particular interest. This can have the benefit of increasing motivation to engage with the assessment as it makes it more meaningful to the student. For instance, coursework could focus on security data analysis and provide students with a choice in the security data set used.

One must be cognisant that choice can be overwhelming for some learners; as such, it is recommended to have a 'default' route, which can be followed by those learners who do not wish to make such a choice. From a marker's perspective, too many choices can also be difficult to manage as they are often less efficient, given the possible variances.

Transparency. It should be clearly communicated to learners which outcomes are being assessed, which is readily achieved if following a constructive alignment approach. It can be made clear to students what is being assessed through, e.g., the use of rubrics, past examples, and opportunities to clarify the assessment specification in an open way which is visible to all students (not just those who think to ask the question). The avenues for support should be clearly outlined, such that students know what steps to take should they become 'stuck'.

Flexibility. Flexibility when reflecting on assessment outcomes could also be considered through the weighting of different outcomes. At the start of the module, the outcomes of the module could be presented to students, and the cohort could provide some input into the outcomes using particular assessment methods. For instance, students could be presented with the range of outcomes of the module and asked which outcomes they would prefer to be assessed in a given way or with a particular weighting.

5.4.3 Artefact

Choice. It may be possible to allow choice in the type of artefact generated to demonstrate the achievement of intended learning outcomes. For instance, if an ILO aims to measure the ability to critically analyse a given security incident, then this could be demonstrated through a presentation, oral assessment, or written report. This can provide choice, which allows each learner to evaluate which format maximises their opportunity to demonstrate the intended objective.

Transparency. It should be transparent to learners which elements of the artefact are being evaluated and how. This can be achieved in several ways, such as by including a clear marking rubric or guide that highlights whether communication, for instance, is being explicitly assessed or whether the style of a presentation is being assessed.

Flexibility. Consider if the related skills, which may be challenging for some individuals, are a core part of the assessment consider (restricted) choice in the format of the artefact.

5.5 A Balancing Act

It is important to recognise that as with everything in assessment, there is the optimal approach and the realistic approach. Whilst a preferred design may be sufficiently well considered in the above aspects, one must also consider the tension between the "optimal" design and the limitations of resources available to deliver such an assessment.

For instance, if a single lecturer is responsible for over 100 students with no marking support, then an assessment which involves presentations to the lecturer is likely to be unfeasible. Consequently, we must aim for inclusive design whilst recognising these limitations.

It should also be noted that there are likely other ways in which the principles of transparency, flexibility, and choice could be applied to cybersecurity assessment, and we encourage other educators to share such practices. Designing an assessment is a reflective process for educators in that one must reflect on how students performed in the assessment once it has been delivered and consider if there are any unintended areas of concern which should be revised before delivery to a new cohort. This, alongside other challenges in designing inclusive cyber assessment, is reflected upon in the next section.

6 Challenges in Designing Inclusive Cyber Assessment

Having identified how educators can embed inclusivity into assessment design, we must also consider that, much like the tension between security and usability, similar trade-offs and challenges may exist in developing inclusive cyber assessment. This section aims to highlight some of these challenges.

6.1 Balancing Authentic and Inclusive Assessment

Designing an assessment which is authentic and inclusive can present a number of challenges; in particular, there can be tension between real-world tasks and environments and engaging learners in the completion of such tasks, which may not be accessible or inclusive for all students. For example, we have discussed the use of specific technologies in cybersecurity that may not be accessible to some students due to their design. Similarly, it can be group work can be a significant stressor for many neurodiverse students (such as those with an autism spectrum condition) to work in teams due to their demands on emotional and energy regulation, though teamwork is a critical element of cybersecurity. In these cases, devising reasonable adjustments may be very difficult.

6.2 Resource Constraints

Resource limitations can further complicate the design of inclusive cyber assessment. Much like implementing hands-on security labs, inclusive design and delivery can mean a higher cost for educators' resources. This is particularly challenging when class sizes are large, which places substantial demand on educators' time and increases the logistical and administrative support which are needed. At times, such an increase in required resources does not match actual resources, meaning we cannot always deliver the assessment we would like to.

6.3 Contract Cheating and Generative AI

An additional challenge is the increasing prevalence of generative AI technology, which can readily create text and other resources that students may utilise in their assessments. Academics are concerned that this could diminish confidence in students' attainment of the desired learning outcomes. The JISC primer on Generative AI proposes three potential strategies for educators: to avoid, to outrun or to embrace and adapt [18].

In some respects, this issue resembles contract cheating. However, it is more readily available due to the current proliferation of free models, unlike contract cheating which incurs a financial cost. Nevertheless, inequalities can still manifest within the context of generative AI. If educators were to permit student use of these tools, some students might be able to afford more advanced models, leading to disparities in access.

Furthermore, generative AI can produce incorrect or incomplete information that students might mistakenly rely upon. Such reliance could adversely impact crucial cybersecurity skills like critique and communication. This risk arises if the generated information is used without critical reflection, as it circumvents the cognitive processes essential for developing comprehension.

Plapp et al. offer valuable insights into these challenges in cybersecurity assessment by examining the effectiveness of a generative AI engine in answering cybersecurity questions [34]. Their research findings led to the following recommendations for enhancing the robustness of cybersecurity assessments against generative AI use: aligning questions or long-format reports with recent knowledge and resources (as models require time to train on up-to-date information); incorporating very recent research and references into questions or reports; and meticulously fact-checking references during marking [34].

7 Conclusion

This chapter has explored the literature to identify key aspects of inclusive assessment design. We have argued that inclusive design is essentially good design that should be aspired to for all students, not solely those with additional support needs. We subsequently reflected upon the core elements of cybersecurity assessments and proposed a process for designing such assessments. An exemplar was presented to illustrate how these steps can be integrated into a cyber educator's practice.

It is crucial to remember that assessment design is inherently a reflective and iterative process. While we contend that good cybersecurity assessment is inherently inclusive, we acknowledge that fully achieving this goal is not straightforward. Nonetheless, it is a worthwhile endeavour. We hope this chapter will encourage other cyber educators to critically examine their own assessment designs.

References

1. Martin, A., Rashid, A., Chivers, H., Danezis, G., Schneider, S., Lupu, E.: The cyber security body of knowledge. University of Bristol, 2021, ch. Introduction to Cybok, version 1.1.0. https://www.cybok.org/
2. Allison, J.: Devising a cyber security management module through integrated course design. J. Furth. High. Educ. **47**(10), 1389–1403 (2023). https://doi.org/10.1080/0309877X.2023.2250729
3. Andress, A.: Surviving Security: How to Integrate People, Process and Technology. Sams White Book Series. Sams (2001). https://books.google.co.uk/books?id=J99QAAAAMAAJ
4. Bain, K., et al.: Inclusive assessment in higher education: what does the literature tells us on how to define and design inclusive assessments? J. Learn. Dev. High. Educ. **27**, 1–23 (2023)
5. Biggs, J.: Enhancing teaching through constructive alignment. High. Educ. **32**, 347–364 (1996). https://doi.org/10.1007/bf00138871
6. Bloom, B., Krathwohl, D.: Taxonomy of Educational Objectives: The Classification of Educational Goals. No. v. 1 in Taxonomy of Educational Objectives: The Classification of Educational Goals. Longmans, Green (1956)
7. Carmines, E.G., Zeller, R.A.: Reliability and Validity Assessment. Sage Publications (1979)
8. Christine Hockings, P.B., Terentjevs, M.: Making a difference-inclusive learning and teaching in higher education through open educational resources. Distance Educ. **33**(2), 237–252 (2012). https://doi.org/10.1080/01587919.2012.692066
9. Transforming assessment and feedback with technology. JISC Innovation Group (2015)

10. Accreditation of higher education programmes (AHEP) (2020). https://www.engc.org.uk/ahep. Accessed 26 July 2024
11. Dawson, J., Thomson, R.: The future cybersecurity workforce: going beyond technical skills for successful cyber performance. Front. Psychol. **9** (2018). https://doi.org/10.3389/fpsyg.2018.00744
12. Downing, S.M.: Validity: on the meaningful interpretation of assessment data. Med. Educ. **37**(9), 830–837 (2003)
13. Du, W.: Seed: hands-on lab exercises for computer security education. IEEE Secur. Priv. **9**(5), 70–73 (2011). https://doi.org/10.1109/MSP.2011.139
14. French, S., Dickerson, A., Mulder, R.: A review of the benefits and drawbacks of high-stakes final examinations in higher education. High. Educ. 1–26 (2023). https://doi.org/10.1007/s10734-023-01148-z
15. Graham, C.M., Lu, Y.: Skills expectations in cybersecurity: semantic network analysis of job advertisements. J. Comput. Inf. Syst. **63**(4), 937–949 (2023). https://doi.org/10.1080/08874417.2022.2115954
16. Hounsell, D.: Contrasting conceptions of essay-writing. Experience Learn. **2**, 106–125 (1997)
17. Irwin, B., Hepplestone, S.: Examining increased flexibility in assessment formats. Assess. Eval. High. Educ. **37**(7), 773–785 (2012)
18. JISC: Generative AI: A primer (2023). https://www.jisc.ac.uk/reports/generative-ai-a-primer. Accessed 22 July 2024
19. Knight, P.T.: The value of a programme-wide approach to assessment. Assess. Eval. High. Educ. **25**(3), 237–251 (2000)
20. Krathwohl, D.R.: A revision of bloom's taxonomy: an overview. Theory Into Pract. **41**(4), 212–218 (2002). https://doi.org/10.1207/s15430421tip4104_2
21. Köhler, D., Meinel, C.: The right tool for the job: contextualization of cybersecurity education and assessment methods. In: Proceedings of the 10th International Conference on Information Systems Security and Privacy - ICISSP, pp. 234–241. INSTICC, SciTePress (2024). https://doi.org/10.5220/0012457900003648
22. LeClair, J., Abraham, S., Shih, L.: An interdisciplinary approach to educating an effective cyber security workforce (2013). https://doi.org/10.1145/2528908.2528923
23. Maguire, J.: Assessing knowledge and skills in forensics with alternative assessment pathways. In: Proceedings of the 6th Conference on Computing Education Practice, pp. 29–32 (2022)
24. Maguire, J., English, R.: Opportunities to fail: using peer-review to support assessment literacy in cyber security. In: Proceedings of the 21st Koli Calling International Conference on Computing Education Research, pp. 1–2 (2021)
25. McEwan, M.P.: The essay as a lens on transition to university: student and staff perceptions of essay writing. Int. J. Teach. Learn. High. Educ. (2016)
26. Messick, S.: Validity of Test Interpretation and Use, 6th edn. Macmillan, New York (1992). https://doi.org/10.1002/j.2333-8504.1990.tb01343.x
27. Mirkovic, J., Peterson, P.A.H.: Class capture-the-flag exercises. In: 2014 USENIX Summit on Gaming, Games, and Gamification in Security Education (3GSE 14). USENIX Association, San Diego, CA (2014). https://www.usenix.org/conference/3gse14/summit-program/presentation/mirkovic
28. Morris, C., Milton, E., Goldstone, R.: Case study: suggesting choice: inclusive assessment processes. High. Educ. Pedagogies **4**(1), 435–447 (2019)
29. Nicol, D.J., Macfarlane-Dick, D.: Formative assessment and self-regulated learning: a model and seven principles of good feedback practice. Stud. High. Educ. **31**(2), 199–218 (2006)

30. Norwich, B.: Education, inclusion and individual differences: recognising and resolving dilemmas. Br. J. Educ. Stud. **50**(4), 482–502 (2002)
31. Orkwis, R., McLane, K.: A curriculum every student can use: design principles for student access. ERIC/OSEP topical brief (1998)
32. O'Neill, G.: It's not fair! students and staff views on the equity of the procedures and outcomes of students' choice of assessment methods. Irish Educ. Stud. **36**(2), 221–236 (2017)
33. Parker, J.: Effective practice learning in social work. Effective Practice Learning in Social Work, pp. 1–176 (2010)
34. Plapp, A., Wu, J., Pan, L., Chen, C., Chua, C., Zhang, J.: Strengthening cyber security education: designing robust assessments for chatgpt-generated answers. In: Kim, D.D., Chen, C. (eds.) Machine Learning for Cyber Security, pp. 17–27. Springer, Singapore (2024)
35. Price, M., Handley, K., Millar, J., O'Donovan, B.: Feedback: all that effort, but what is the effect? Assess. Eval. High. Educ. **35**(3), 277–289 (2010)
36. Prickett, T., et al.: Challenges and Opportunities of Teaching Cybersecurity in UK University Computing Programmes, pp. 1–35. Springer, Cham (2023). https://doi.org/10.1007/978-3-031-24216-8_1
37. Rose, D.: Universal design for learning. J. Spec. Educ. Technol. **15**(4), 47–51 (2000)
38. Sambell, K., McDowell, L., Brown, S.: "But is it fair?": an exploratory study of student perceptions of the consequential validity of assessment. Stud. Educ. Eval. **23**(4), 349–371 (1997)
39. Schuwirth, L., Van der Vleuten, C., Durning, S.: What programmatic assessment in medical education can learn from healthcare. Perspect. Med. Educ. **6**, 211–215 (2017)
40. St. Lucia, A.T.U.o.Q.: Flexible assessment in a business course (2001). http://www.tedi.uq.edu.au/largeclasses/pdfs/CaseStudy-02_ChapAsafu.pdf
41. Tai, J., Ajjawi, R., Umarova, A.: How do students experience inclusive assessment? A critical review of contemporary literature. Int. J. Inclusive Educ. 1–18 (2021)
42. Taras, M.: Using assessment for learning and learning from assessment. Assess. Eval. High. Educ. **27**(6), 501–510 (2002)
43. Van Der Vleuten, C.P., Schuwirth, L., Driessen, E., Govaerts, M., Heeneman, S.: Twelve tips for programmatic assessment. Med. Teach. **37**(7), 641–646 (2015)
44. Vykopal, J., Švábenský, V., Chang, E.C.: Benefits and pitfalls of using capture the flag games in university courses. In: Proceedings of the 51st ACM Technical Symposium on Computer Science Education, SIGCSE 2020, pp. 752–758. Association for Computing Machinery, New York (2020). https://doi.org/10.1145/3328778.3366893
45. Wilde, A.: Equality, diversity, and inclusion in the CS curriculum. In: Proceedings of the 2023 Conference on United Kingdom & Ireland Computing Education Research, UKICER 2023. Association for Computing Machinery, New York (2023). https://doi.org/10.1145/3610969.3611125
46. Wilde, A., Dix, A., Evans, C., Vasilchenko, A., Maguire, J., Snow, S.: Towards a taxonomy of video for HCI education, chap. 2, pp. 31–46 (2019). https://doi.org/10.36006/16184-02
47. Wilde, A., Sanders, B.: Transforming postgraduate cybersecurity education: a cybersecurity body of knowledge case study, pp. 1474–1479 (2024). https://doi.org/10.21125/inted.2024.0432

Cyber Security Unplugged: Teaching Security Management Through Immersive Gameplay

Alan Mills[1](✉), Ian Caple[1], Phil Legg[1], Sophie Fenn[2], and Aida Abzhaparova[1]

[1] University of the West of England, Bristol, UK
alan.mills@uwe.ac.uk

[2] Pate's Grammar School, Cheltenham, UK

Abstract. Cyber security education and the online safety of young people is key on the government agenda. A variety of initiatives are designed to raise awareness about the threats and safeguards related to children and their use of technology. Furthermore, cyber security incidents are frequently impacting young people by directly disrupting the services they rely on, from schools to social media. As a result of a vast array of national, regional and local government funded programmes, young people are becoming increasingly aware of the importance of cyber security, both as an a societal need but also as a potential career path. However, what remains is still a gap between the need for cyber education and the diverse range of careers in this field. This chapter presents an innovative pedagogic and gamified experience for teaching fundamental security management concepts, that differs from typical learning environments that focus on technical skills. The game was co-developed with school teachers and designed for teachers and educators to use in a variety of settings, such as schools, HE, and business. This game has been successful in demystifying the field of cyber security for new players, explaining the entry points into the field, and demonstrating the wide-range of skills and behaviours required to pursue careers in cyber security. By drawing from the successful application of this game, the chapter highlights the importance of using immersive, offline, and non-technical ("unplugged") pedagogies. Furthermore, it demonstrates how educators can explain the complexities of cyber security and entice young people to further explore a multitude of roles and careers in this field.

1 Introduction

This chapter presents aspects of cyber-security education related to security management and decision making in cyber security crisis scenarios. The current approaches to cyber education predominately focus on technical skills development, such as Network Traffic Analysis (NTA), reverse engineering, and penetration testing. In contrast this chapter proposes and demonstrates the importance of non-technical entry points into a field where a lack of such skills can often be perceived as "barriers to entry". Following the recent reporting of the challenges and skills gaps within cyber security, this chapter argues in favour of adding non-technical pedagogies into cyber security through an offline, scalable Incident Response Management (IRM) game[1].

[1] https://github.com/uwe-cyber/teachersworkshop2024/blob/main/lesson-IRM.md.

P. Legg et al. (Eds.): CSE 2024, LNNS 1213, pp. 213–234, 2024.
https://doi.org/10.1007/978-3-031-77524-6_12

In this chapter we present our findings and feedback from utilising the IRM game to highlight the importance of teaching cyber security through non-technical means across a diverse group of learners. The chapter starts with detailing motivation and background to the development of the game. We then discuss the iterative and feedback-driven process that informed the development of the game and its key mechanisms. The chapter benefits from a pedagogic case detailing implementation of the game within a classroom. Here we provide analytical reflections on how the game meets the educational needs, as well as helps to develop key professional skills among learners.

2 Background and Motivation

2.1 Industry and Academia

The skills gap in cyber security is an ongoing academic, industry and governmental concern, with the Department for Science, Innovation and Technology (DSIT) reporting on the ongoing cyber security skills gap. The recent report (DSIT, 2023, p. 2) highlights issues around technical skills gaps, with cyber security governance and risk management being the second highest gap. The report also states that less than half of staff in the cyber sector had previously worked in cyber roles elsewhere (ibid), highlighting the need for education and upskilling in industry. There is also increasing demand from government and industry to increase the diversity of the cyber security workforce, with the DSIT reporting (ibid) which is highlighted an urgent need to encourage more females into cyber security. This supports the aims of this chapter in ensuring that cyber security education is as accessible as possible from the outset and extending from Key Stage (KS) 3 to Continual Professional Development (CPD). These industry-based pressures have also been the subject for academic research and curriculum-related discussions in both education and higher education.

In their work Goupil et al. (2022) analysed data from cyber security jobs advertisements and academic curricula to identify areas of need, and areas of "undersupply". They find that security management has one of the highest skills gaps as evidenced by Fig. 1. This highlights the industrial demand and skills gap for security management, which is in direct contract to the oversupply for areas like security certification.

In a similar study Ramezan (2023) looked at the requirements of 935 cyber security-based jobs, breaking down requirements to look at: education, experience, certifications, (security) clearance, and programming skills. They found that less than half (48.3%) of the advertised jobs mentioned certificates and only 21.1% had an actual requirement, instead of preference (ibid, p. 99). Similarly, only 24.7% listed a programming language in the job advert. These findings add to those shown by Goupil et al. (2022) and reinforce their concern that cyber security problems cannot be solved by technical skills alone and that cyber security education needs to provide "stronger offerings" in non-technical domains and incident response. In other words, attracting students beyond those who are actively engaged and motivated by technical aspects of computer science.

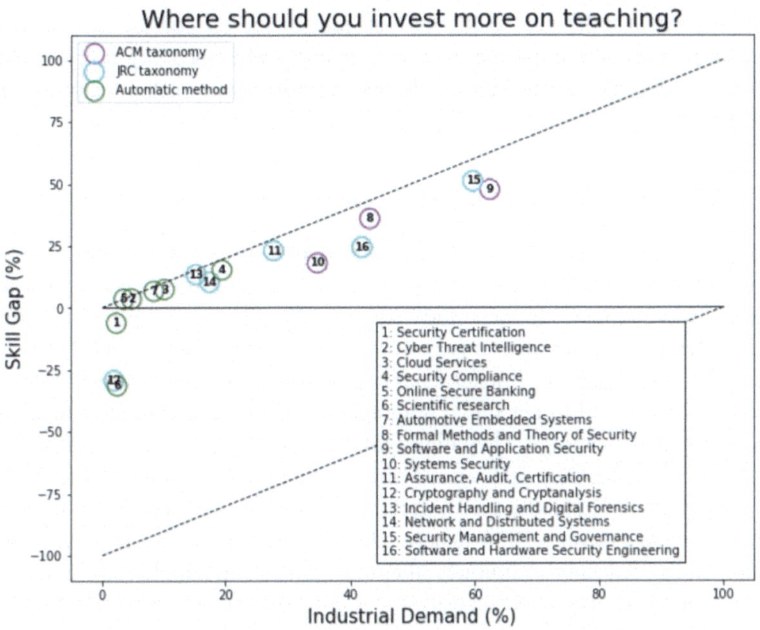

Fig. 1. Summary and prioritization of identified skill (Goupil, F et al., 2022)

Existing platforms like TryHackMe[2], HackTheBox[3] and ImmersiveLabs[4] lend themselves to a historical focus on security certifications and areas like penetration testing. This has in part left a significant skills gap in areas like security management and which must be addressed in a way that scales across both academia and industry. We consider the following non-technical, professional skills to be key requirements to address these areas of undersupply: decision making; collaboration; communication and advocacy; the ability to work under pressure (or time constraints); and the ability to maintain an analytical process throughout. These skills are fundamental to successful careers in cyber security. Importantly, the professional skills under consideration and their lack contributes to necessary industry shortages, such as security and risk management. Furthermore, facilitating a safe learning environment where learners can feel comfortable in exploring cyber security challenges, and can allow them to make mistakes and recover in a non-judgement way, is essential in developing confident and resilient learners.

The teaching of these non-technical, professional skills also has an impact beyond cyber security specific domains. In their work Walkington et al. (2018) focus on a "capabilities approach" to education highlighting that educators should think beyond domain specific considerations to what learners will gain from their education as part of their boarded development. They discuss the importance of decision making skills amongst others, with communication and team working being great contributing factors

---

2 https://tryhackme.com/.

3 https://www.hackthebox.com/.

4 https://www.immersivelabs.com/.

to developing social skills and "resilience that stems from self-confidence to self-belief". This in turn prompts the importance of recognising and embedding non-technical and professional skills in order to help apply and amplify some of the technical skills of cyber security.

2.2 Serious Games and Game Based Learning

The use of Serious Games (SGs) and Game Based Learning (GBL) has already been explored within academia for multiple subject areas at all levels including Continuous Professional Development. In their study Angafor, Yevseyeva, and He, (2020) reviewed multiple tabletop exercises with a focus on cybersecurity. They found that the use of SGs and GBL had a positive impact, enabling acquisition of "essential skills" but that they should be designed to teach and demonstrate technical and nontechnical skills to address current shortcomings. They highlight the lack of staff with key skills, including communication and decision making and discuss the requirement for non-technical staff to "understand the threat landscape" (Angafor, Yevseyeva, and He, 2020). In their study Winger, Ellis, and Glover, (2024) teach cyber strategy through crisis simulation and found that "active learning techniques like mini-games and simulations hold particular promise for demystifying cybersecurity for non-technical students". They discuss the advantages of using such methods to engage students in subjects which are often taught at abstract levels and the subsequent interest and positive attitudes show by students, even those who lacked cyber specific knowledge.

In their report Hallet, and Shreeve (2022) map existing games onto the Cyber Security Body of Knowledge (CyBOK[5]) and analysed both the game content and gameplay. They looked at 3 existing games; Decisions & Disruptions (D&D), [dox3d!], and Elevation of Privilege (EoP). Of these games only D&D was found to have a focus on areas such as risk management. While both [dox3d!] and EoP use card mechanics similar to those used by IRM, they focus on putting players into the role of hackers or software engineers retrospectively. In their review of [dox3d!] the authors found that while it did provide engagement and enthuse players, "it did not appear to promote cyber security thought or reflection without supplemental prompting" (Hallet, and Shreeve, 2022). With EoP they found that the game at times required players to already be very familiar with the underlying material. This therefore made it unsuitable for beginners or those without existing cyber security knowledge, leading in some cases to confusion and subsequently a swift loss of interest when playing the game (Hallet, and Shreeve, 2022). In D&D players focus on securing a utility company which is represented by LEGO. The game itself lasts 4 rounds, under the direction of a games master and players must make decisions each round on how they will spend their budget to secure their utility company. This has been shown to address key knowledge areas within risk management but was largely aimed at Higher Education (HE) audiences or industry professionals and would be difficult to scale due to the physical requirements and need for a games master to oversee each game (in a similar style to Dungeons & Dragons). A second version of the game, D&D2: Decide, Harder Shreeve, et al (2023), has since been released which provides a digital

[5] https://www.cybok.org/.

offering and User Interface as well including more complex threat relationships and additional randomisation within the game dynamics.

Riskio is a similar tabletop game utilising a game board and playing cards, which is presented in the work by Hart, et al. (2020). As with D&D(2) the game makes use of a games master and has been tested with university and industry players. The objective of the game is for players to defend against drawn attacks, with the game master themselves acting as an arbitrator and single point of knowledge or truth "the game master explains which among the played Defence cards was effective and why the others were not" (Hart, et al., 2020). Both attack and defence cards are high level, for example "Network Monitoring" card which covers the use of "tools" for monitoring key events. Similarly in the attack deck there are different "Spoofing Attacks", email, passwords and websites. While these concepts work at the higher level of cyber security education, it does not provide explicit mapping to educational frameworks for KS3 – 5 students and the game and the requirement for a game master makes it difficult to scale inside a classroom setting.

Our game – Incident Response Management (IRM) builds on these concepts, providing a card-based mechanic that allows for the same knowledge areas to be covered through a scalable, non-technical solution that can be independently played by multiple groups at once in a classroom setting. The IRM game is designed with educational goals at its core, providing a versatile tool for teaching essential knowledge, skills, and competencies in cybersecurity and computer networking. Its flexible design allows it to be tailored to specific learning objectives across Key Stages 3, 4, and 5, and is well aligned with the National Curriculum and current GCSE and A-level specifications for Computer Science (see *Concepts, Skills and Competency mapping*). Importantly IRM can be mapped to educational and industry frameworks, such as CyBOK, NIST cybersecurity framework[6], Lockheed Martin Cyber Kill Chain[7] as well as the UK Cyber Security Council Career Framework[8]. This makes it easy to utilise IRM within existing lesson plans at multiple education levels and even as CPD, something that was achieved due to the collaborative design and development between teachers, academia and industry.

3 Game Design

The game was created as a result of ongoing teacher training and outreach between The University of the West of England (UWE) Bristol and schools within the South-West of England. In January 2023, during one of the training workshops, teachers were tasked to develop a lesson plan which could focus on careers within cyber security. A common theme with all the teachers was that technical roles and "hacking" were already well known by their students, to the point where many pupils considered these the only job roles within cyber security. The teachers' discussion at the workshop reinforced an already established perception that a lack of technical skills was a barrier to entry for the cyber security industry. This in turn could lead to many students losing initial interest and subsequently dismissing cyber security as a visible career path with versatile job roles.

[6] https://www.nist.gov/cyberframework.

[7] -https://www.lockheedmartin.com/en-us/capabilities/cyber/cyber-kill-chain.html.

[8] https://www.ukcybersecuritycouncil.org.uk/careers-and-learning/cyber-career-framework/.

The teachers also highlighted how it would be useful to have an "unplugged" lesson. In other words, something that could be delivered without any technology, in part to provide a respite from the use of computers and to avoid reliance on them, especially as the setup can often take up valuable teaching time or significantly interrupt learning if technology fails.

Informed by the workshop discussion and activity, IRM was co-created as a tabletop card game that brings together elements of security and risk management with incident response. Informed by the outcomes of collaboration between industry, education and academia, the game allows players to gain insight into asset and risk management within cyber security. The game was designed around the concept of learners, not players (see *Pedagogic Case*). In addition, rather than emphasising the technical aspects of cyber security, the game prioritises the importance of professional skills, highlighting to the learners that there are roles within cyber security which do not require them to have significant technical skills. Complex technical concepts are learnt through practicing professional skills such as: decision making, collaboration, working under pressure, risk management, and an analytical process. Ultimately the game is designed to showcase the non-technical careers and entry-points within cyber security. Highlighting to the students that if they can utilise professional skills such as collaboration, making important decisions and maintaining an analytical process throughout, there is a role in cyber security for them. The game has had two iterations and has evolved over time as a result of teachers' and learners' feedback. The next section details step-by-step the evolution of the game including reflective discussion on aspects that have been introduced to achieve maximum learning engagement and impact.

3.1 Initial Game Design

In its initial form the game was composed of three card types; Contracts, Blue Team assets and Threats, as shown in Fig. 2 (from top to bottom).

A game starts with learners divided into small groups, who are initially provided with starting capital and reputation. This starting capital could be used to purchase assets at the start of the game. A threat card would then be randomly drawn. If the learners had the assets in place to mitigate the threat, then they would move onto the next game phase, gaining a contract. If they could not (fully) mitigate the threat, then they would not be able to do this as they would be dealing with the impact of the successful attack. In the contract phase the learners would look at the available contract cards and try to find a contract they could acquire. Contracts would all have different requirements, based around the assets that the learners had. For example, the Multimedia contract (see Fig. 3) requires the learners have a Firewall and either an Intrusion Detection or Prevention System (IDS/IPS) in place, with requirements varying between contracts. After the contract phase that turn has ended and they start a new turn again from the initial Asset phase, with the discussion had during their contract phase feeding into and potentially informing their choices around which asset to purchase. Figure 4 provides a breakdown of the game phases and flow.

Importantly contracts will also have penalties associated with them. For instance, in the case of the Multimedia contract the learners will lose more than they gained if a successful Data Exfiltration attack occurs. As such the groups need to balance their asset

Fig. 2. Original Incident Response Management game card deck

purchases carefully. Do they purchase assets to meet requirements for new contracts? Or do they purchase assets to reinforce their existing defences?

As the game progresses and they purchase more assets new contracts become available, allowing learners to acquire more capital to re-invest and develop more assets, in doing so revealing the relationship between business management and cyber resilience. Should the learners survive enough threats and build a Security Operations Centre (SOC) they effectively win the game. Games can also be tied to a number of rounds with a winning team being decided based on a combination of capital, reputation and available assets or contracts.

In February 2024, the original IRM game was presented as part of the CyberWEST[9] (West of England Schools and Teachers) project which involved multiple Universities

[9] https://uwe-cyber.github.io/cyberwest/.

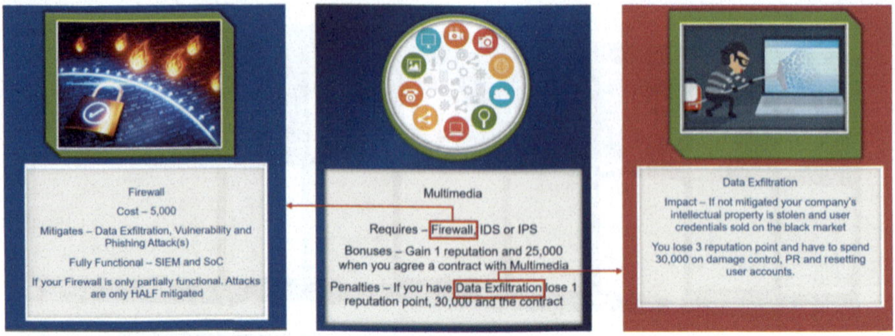

Fig. 3. Asset (left), Contract (centre) and Threat (left) cards

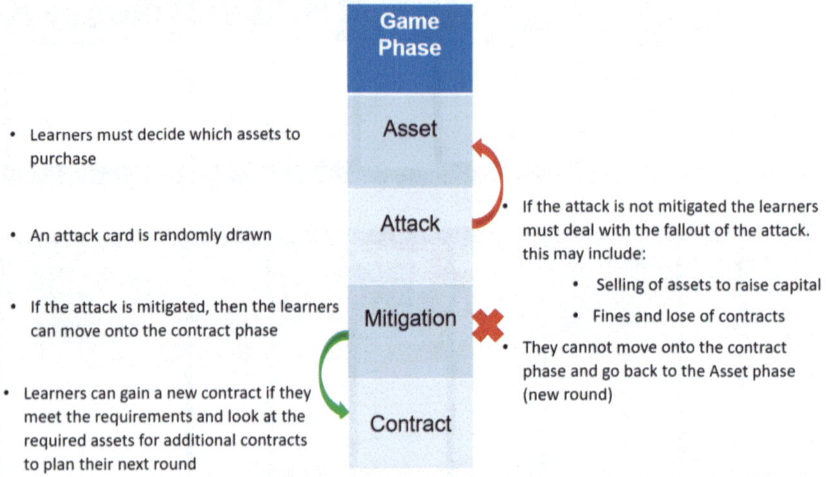

Fig. 4. Game phases and flow

within the South-West region working together as part of outreach and collaboration with regional schools. It formed part of our activity offerings, and we asked all the teachers for verbal feedback during and anonymous feedback at the end of the event. Overall, the IRM game was one of the most well received activities with positive rating and feedback such as:

> *"Loved the physical card game, more than the digital. Physical card game can get some really good discussions going. Definitely will use."*
>
> *"Brilliant. Learners use technical language to justify choices. Learner can research different forms of mitigation."*

We also showcased the game with a cross-discipline group of HE students at UWE Bristol. While no direct feedback was provided, students did make social media posts and comments via LinkedIn about the session itself. Below we present some of their feedback:

"The simulation taught us a vital lesson: even with the most comprehensive threat prevention measures [...] unforeseen threat could (and most probably will) still arise."

"Not only did it help us understand the complexities of real-world scenarios but it also gave us a taste of the stress [pressures] that comes with them."

Overall, the feedback highlighted that teachers found the gamification of this possibly "dry" subject, the discussion it encouraged, and team based play styles to be success factors. Whilst (HE) students enjoyed the immersion itself, which helped provide a sense of realism around risk management, handling threats, and the pressures this can introduce when working in industry.

It is important to note that both teachers' and HE students' feedback informed our iterative design process, and fed into the next version of IRM, Red vs Blue Three particularly significant points were the need to:

- Add additional complexity and competitiveness to entice learners and keep their engagement throughout the game.
- Provide richer and informative descriptions of each card, and ensure realistic imagery while acknowledging the importance of accessibility, and Equality and Diversity.
- Extend and reflect the range of job roles within cyber security.

In the following section, we detail how the game has evolved in response to the teachers' and students' feedback.

3.2 Iterative Game Design

In addition to the three points above, the team also sought to approach the further design of the game in a holistic, reflective and realistic manner. This included:

- Customisation of the Threat/Attack and Asset Decks to ensure accurate and reflective mapping to Key Stage 3 and Key Stage 4 Computer Science curriculum in England (curricula delivered to the students between the ages of 12–16 years old).
- Redesign of the game card graphical representation with the focus on accessibility in terms of colour of the cards.
- Addition of Chief Information Security Officer (CISO) cards to fully reflect organisational context, structure and responsibilities at the time of cyber incidents.

Introduction of a Playable Red Team

The original version of the game was based solely around learners playing Blue Team roles, as such threats are randomly drawn from the deck. This adds an element of randomisation to what they might face but does mean there is no intelligent targeting as might occur in the real world. The addition of the Red Team Asset cards and mapping them to the Attack cards had a significant impact on the game play and learner experience. No longer were learners "playing against the deck", they now had an adversary across the table from them. This helped to address mapping elements that were missing in the original, such as the Reconnaissance phase of the Lockheed Martin Cyber Kill Chain or more advanced industry frameworks, such as the NIST Cybersecurity Framework (CSF), for use as CPD as evidenced in (see *Concepts, skills and competency mapping*).

Cyber offensive assets such as Command and Control (C2) Servers, Rubber Ducky – Bad USB devices and Vulnerability Scanners were created along with industry standard job roles such as OSINT (Open Source Intelligence) Officer, Penetration Specialist and Social Engineer. This helped feed into the expansion of career paths within the game and enabled learners to grow their aware-ness of the Red Team side of the industry. These assets were then mapped to existing and newly created Attack cards, such as ransomware and insider threat (see Fig. 8).

Visuals, Representation and Accessibility

Figure 5 illustrates how the graphical presentation of all cards within the game evolved to address feedback such as asset descriptions, realistic imagery for job role focused assets, and diversity within these assets. This diversity and awareness of differing jobs roles was designed to empower learners of all backgrounds to feel they have a role within cyber security, whilst retaining the core mechanics of the original game.

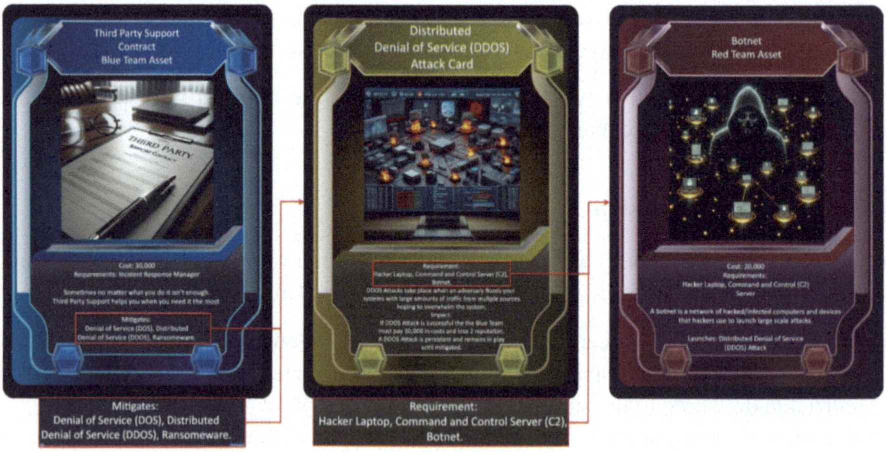

Fig. 5. Redesigned SOC Analyst Card

Expansion of Cyber Security Career Paths

To address the lack of job roles in the original game, the number of cyber security job roles within the Blue Asset deck was expanded from 1 to 3 (see Fig. 6). These now included descriptions and definitions of these roles to enable learners to gain an understanding of the roles, and their place and function within the cyber security environment. These expanded career paths mirrored the real-world organisational structure of a cyber security team, with each path providing the Blue Team with benefits related to their job role. The differing career paths and the way in which they function are designed to give the learners insight into the dynamics and relationships of not just the roles in a corporate context, but also the roles they take in responding to a cyber incident.

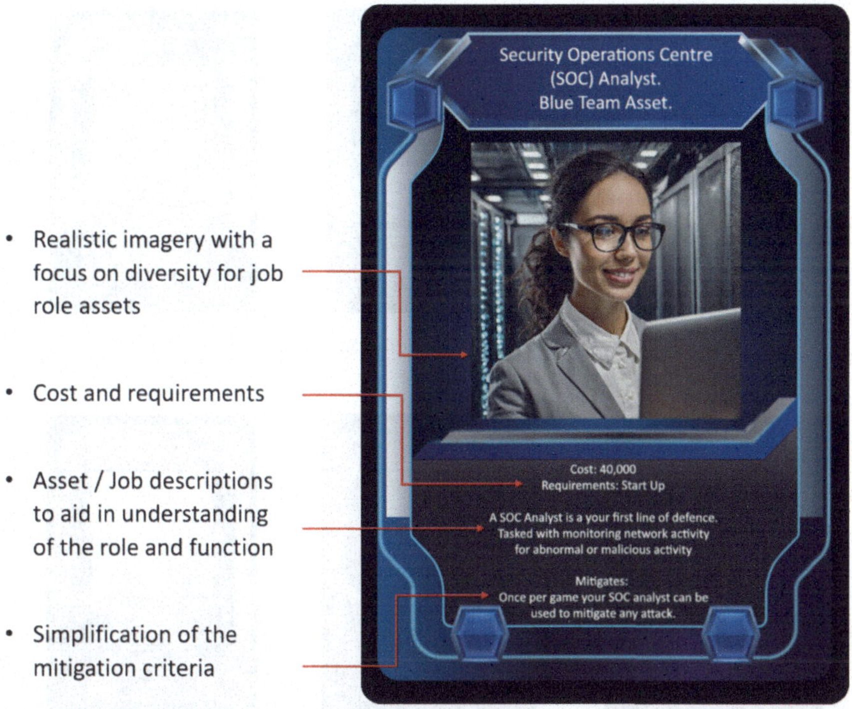

Fig. 6. The expanded career path assets added to IRM: Red vs Blue

Building upon this further a new card type was added. The purple CISO (Chief Information Security Officer) card. These CISO cards grant a variety of in game benefits to players, based upon the background of each CISO, such as automatic attack mitigations, asset or capital bonuses. The CISO cards all have varied career trajectories, based upon real world entry points, highlighting the diverse routes into cyber security. With the expansion of career paths and human assets, a keen focus was placed upon inclusion, diversity (DSIT, 2023, P. 17) and contributing to the government and cyber security commitment to equality, diversity and inclusivity (Fig. 7).

4 Concepts, Skills and Competency Mapping

4.1 Education: Explaining Complex Concepts Through Game

It is crucial to emphasise that the game is not designed as a form of entertainment but as a powerful educational tool. Aligning the game with curriculum standards and learning outcomes ensures that students understand its purpose and the valuable skills they are developing. During our discussions and collaboration with school teachers it became apparent that constraints around teaching schedules and curriculum limited their usage of external resources. As useful as something like the IRM game might be, if it could not be mapped to what they needed to teach, they would often struggle to find time for

Fig. 7. A selection of CISO cards

Fig. 8. Blue Team Assets (left), Attacks (centre), and Red Team Assets (right)

it in their delivery. As such we worked on putting together lesson plans which would provide mapping to specifications such as OCR and AQA. This allows teachers to easily incorporate the IRM game into lessons which are timetabled for relevant topics, for example Fig. 9 shows the GCSE specification for Computer Science taken from OCR. There are clear and direct mappings between this specification and the IRM game, with cards such as ransomware and phishing attacks directly mapping to the forms of attack they need to be covered. The same is also true for some of the prevention methods, with firewall and penetration testing cards present within the game. Importantly, the game itself covers key points within the Guidance column. By playing the game students learn about the different threats and attacks, as well as how they are used. At the same time the different assets that can mitigate these attacks and what different assets can and can't mitigate are covered. There are some elements around attack purpose and how assets limit attacks which are not fully covered. However, to address this a reflective exercise in partnerships with school teachers has taken place and key outcomes are detailed in the *Discussions* section.

The game also provides a similar mapping for students across multiple educational boundaries through the CyBOK Risk Management and Governance Knowledge Tree (v1.1.1). As discussed in *Initial Game Design* when purchasing assets the learners will need to decide if they are focusing on assets for specific contracts or assets that protect them against specific threats. This means they will instinctively be making assessments

1.4 – Network security	
Sub topic	**Guidance**
1.4.1 Threats to computer systems and networks	
☐ Forms of attack: ○ Malware ○ Social engineering, e.g. phishing, people as the 'weak point' ○ Brute-force attacks ○ Denial of service attacks ○ Data interception and theft ○ The concept of SQL injection	**Required** ✓ Threats posed to devices/systems ✓ Knowledge/principles of each form of attack including: ▪ How the attack is used ▪ The purpose of the attack
1.4.2 Identifying and preventing vulnerabilities	
☐ Common prevention methods: ○ Penetration testing ○ Anti-malware software ○ Firewalls ○ User access levels ○ Passwords ○ Encryption ○ Physical security	**Required** ✓ Understanding of how to limit the threats posed in 1.4.1 ✓ Understanding of methods to remove vulnerabilities ✓ Knowledge/principles of each prevention method: ▪ What each prevention method may limit/prevent ▪ How it limits the attack

Fig. 9. OCR Computer Science GCSE Specifications

and decisions on the assets they wish to purchase based around the elements of risk and their perceived importance, all of which are topic areas within this knowledge tree and key professional skills we want to focus on and help them develop. When playing the game with students from Higher Education (HE) it is this mapping and the link to crucial cyber security skills required which are highlighted.

4.2 Skills: Exploring Career Pathways Through Play

A key aim during the development of the game was to ensure it was engaging and of interest to students, being able to link it to relevant careers was vital. By highlighting to students the associated skills the game re-enforced and associated careers we could use this to "hook" initial student engagement by tying it to employability and long term career prospects. To that end we have mapped the different game phases to cyber security skills that they reinforce as taken from the Prospects website[10]. Figure 10 shows some of the key skills listed for the cyber security analyst job role and where they would align within the IRM game flow.

As learners move through the game phases they will need to deal with the impact of cyber-attacks and collaborate with others to decide on which assets they will purchase, inviting discussion around a focus on either specific contracts or defensive capabilities. This in turn will feed into their understanding and working knowledge of the different security technologies they have available (their assets) and potentially those they have ignored. Importantly it also helps to simulate the stress and working pressures of dealing with a cyber security incident, as discussed in *Background and Motivation* in an environment that allows the learners to fail safely.

[10] https://www.prospects.ac.uk/job-profiles/cyber-security-analyst.

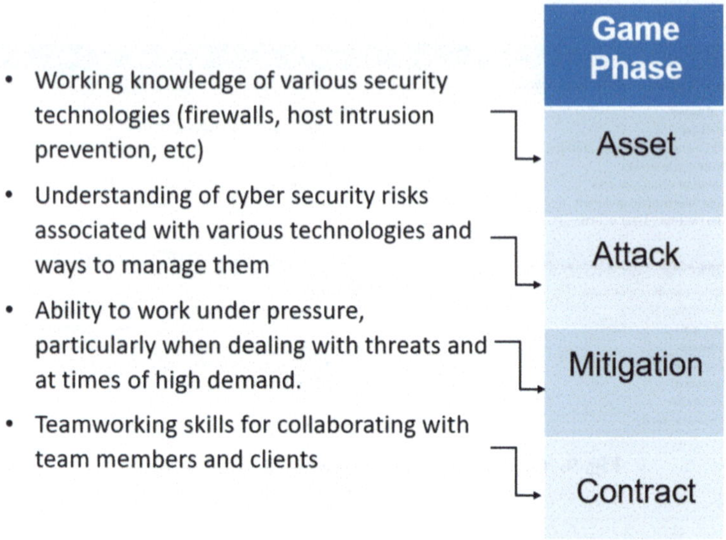

Fig. 10. Mapping of career skills (from prospects) to game phases

4.3 Competency: Industry and Continuous Professional Development

Mapping IRM to industry standards for use as a professional CPD tool was an integral part of the iterative development of the game. The inclusion 2 playable teams and Red Team Asset cards meant more advanced cyber security concepts were introduced such as adversarial behaviour and Tactics, Techniques and Procedures (TTPs). These advanced concepts allowed IRM to be scalable across the learning spectrum from secondary education through higher education and into industry CPD.

In the original game version, the relationship between Blue Team Assets, Contracts and Attack cards was well established, highlighting to players the technologies required on the defensive side. However, the link between technology and offensive cyber security was missing. By incorporating the Red Team Assets into the game that relationship gets played out in full. This enables advanced learners, and in particular industry professionals, to utilise the game for strategizing the development and deployment of their assets. Blue Team players can review Red Team assets in play and attempt to predict what type of attacks the Red Team can execute. This mirrors cyber threat intelligence modelling of adversarial TTP's. Though these advanced learning objectives may not be relevant for Key Stage 3 or 4 students, it enables IRM to scale through the educational ladder up to industry. This included industry standards such as the NIST Cyber Security Core Framework 2.0 and the Lockhead Martin – Cyber Kill Chain (see Fig. 11).

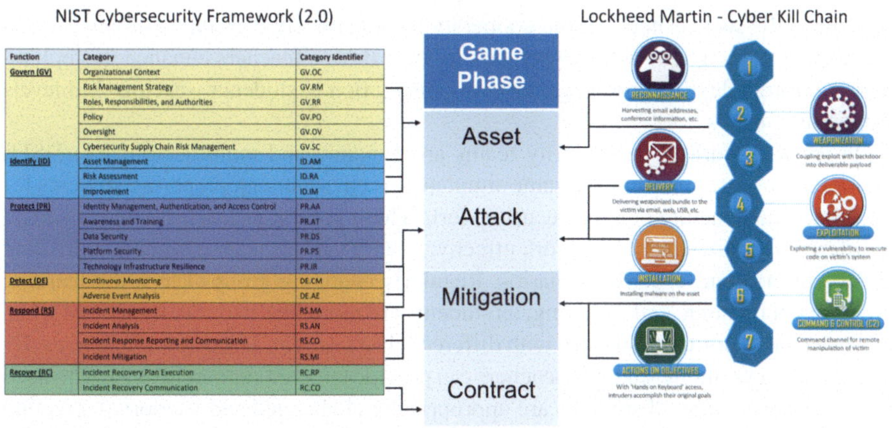

Fig. 11. Mapping of IRM to the NIST Cyber Security Framework 2.0 and Lockhead Martin – Cyber Kill

5 Pedagogic Case

This section explores the integration of the IRM game into educational curricula to enhance student engagement and practical understanding of cybersecurity. The IRM game was tested in Pate's Grammar school in March 2024. The IRM game transitions students from passive learning to active engagement, offering a dynamic and practical approach to understanding cyber security landscape, including cyber threats and mitigation methods, and career pathways.

5.1 Learners Not Players

When integrating the IRM game into the curriculum, the emphasis is on transitioning from passive learning to active engagement. Introducing students to cyber threats and mitigation methods in a dynamic and practical way can be challenging. Often, students become merely recipients of information, needing more context about how these threats manifest in the real world. The game changes this dynamic by making students active participants in the learning process, fostering a deeper and more practical understanding of cybersecurity. As active learners, students make decisions, solve problems, respond to challenges, and face the consequences of those decisions in a simulated environment. This approach significantly boosts student engagement and motivation.

5.2 Collaboration and Competition

A crucial element of the game-based learning approach is the introduction of collaboration and competition, particularly in the IRM: Red vs Blue evolution of the game. In the real world, cybersecurity is a collaborative effort requiring effective teamwork and communication. The game reflects this by allowing students to work together in small groups, making joint decisions. This collaborative environment helps students develop essential skills such as negotiation, cooperation, and collective problem-solving, which are indispensable in real-world cybersecurity operations. Teachers enhance this collaboration through flexible groupings, tailoring instruction to students' varying skill levels and learning needs.

In the real world, cybersecurity teams are dynamic and multidisciplinary (DSIT, 2023). Flexible groupings in the game simulate this environment, preparing students for the collaborative nature of cybersecurity work. These groupings teach students to adapt to different team dynamics and work effectively with various team members, mirroring real-world incident response scenarios. Regularly changing groups keeps the learning environment dynamic and engaging, as students are more likely to stay motivated when they have the opportunity to work with different classmates and face new challenges. By periodically regrouping students, teachers can provide targeted support to those who need it most, ensuring that all students are appropriately challenged and supported, creating a more inclusive learning environment (Education Endowment Foundation, 2020).

Research has shown that collaboration in small groups is particularly effective in encouraging diversity and helping female learners build confidence and enjoy learning more (National Centre for Computing Education, 2021). By working in small, diverse groups, female students feel more included and supported, helping to break down gender barriers and stereotypes in computer science. This collaborative and supportive environment fosters a sense of belonging and empowerment among female students, encouraging active participation. In general, mixed-ability groups are particularly effective, as more knowledgeable students can help teach their peers, reinforcing their own understanding while supporting others. This approach aligns with Vygotsky's Zone of Proximal Development, ensuring that students are always working within their capability while being challenged to grow (Vygotsky, 1978).

5.3 Knowledge Skills and Competencies

Core Knowledge at Secondary Education

At KS3, the National Curriculum emphasises foundational knowledge of computer systems, including understanding hardware, software components, and their interactions within networks. The IRM game introduces students to common network threats and core hardware components through practical scenarios, fostering discussions around essential concepts. Teachers can leverage the game to introduce new concepts at a high level, focusing on guiding students through a few carefully selected key terms and reflecting on the strategies they used during gameplay. This prepares students for GCSE studies and can help to spark an early interest in cybersecurity by challenging the stereotype that all roles in the field are purely technical.

At KS4, students develop a foundational understanding of network architecture, protocols, and common cyber threats and mitigation measures. They also explore the impact of new technologies from ethical, social, and environmental perspectives. The IRM game challenges students with scenarios requiring them to design and manage network configurations, perform risk assessments, and develop incident response strategies. This approach builds their technical knowledge while enhancing their analytical thinking, problem-solving, and strategic planning skills.

Teachers will want to focus on embedding key terminology at this level and ensuring students become familiar with this terminology. To achieve this, educators can employ the concept of semantic waves, which involves moving between abstract concepts and concrete applications to deepen understanding (Curzon et al., 2018). Initially, teachers introduce key cybersecurity terms and concepts at an abstract level, explaining their importance and theoretical underpinnings. Following this introduction, the IRM game is used to embed and consolidate learning by allowing students to apply these terms in practical, real-world scenarios. This hands-on experience helps students internalise the terminology, making it more meaningful and memorable. After the gameplay, teachers bring students back to the abstract concepts, which are now enriched with practical understanding. This cyclical process of moving between the abstract and concrete helps solidify students' grasp of key terms and concepts, ensuring they are well-prepared for both exams and real-world applications. Through this approach, the IRM game reinforces technical knowledge and strengthens students' ability to articulate and apply cybersecurity terminology effectively.

At KS5, A-Level students deepen their understanding of network topologies, protocols, encryption, and security measures. The IRM game simulates complex network environments and sophisticated cyber-attacks, requiring students to apply critical thinking to solve security problems. Through these scenarios, students learn to demonstrate professional behaviours and adhere to cybersecurity practices and guidelines, preparing them for higher education or careers in cybersecurity.

Skills

The IRM game fosters the development of both technical and professional skills. Teachers employ guided discovery, allowing students to explore concepts independently but under supervision. This method ensures students benefit from autonomy while staying focused on learning objectives. For KS3 students, the game introduces strategic thinking and core technical knowledge, laying a foundation for more advanced studies. At KS4, the game enhances critical skills like threat analysis, risk management, and decision-making. Students also develop professional skills such as analytical thinking and problem-solving. For KS5, the game focuses on professional-level skills, including complex problem-solving and strategic planning, mirroring real-world cybersecurity challenges.

Competencies

The IRM game is also designed to instil professional behaviours such as decision-making under pressure, ethical considerations, and adaptability to changing scenarios. At each key stage, the game incorporates scenarios that require students to navigate ethical dilemmas and stay updated with the latest security trends. This aligns with the current focus on developing students' analytical, problem-solving, and computational thinking skills in a cybersecurity context. Teachers can reinforce these competencies through debriefing sessions, encouraging students to reflect on their in-game decisions and outcomes. This reflection helps students understand the relevance of the skills and competencies they develop during gameplay.

Feedback Mechanisms

The IRM game is an excellent tool for providing various forms of feedback, which is crucial for enhancing learning outcomes in computing education. The game offers many opportunities for formative assessment, which aligns with current computing pedagogy and supports effective teaching and learning strategies.

Formative Assessment

The IRM game can be used for formative assessment by providing immediate feedback on students' actions and decisions during gameplay. As students respond to scenarios faced during gameplay, the cards provide real-time feedback about which mitigation strategies are effective against threats faced, as well as help students to identify misconceptions and areas for improvement in understanding key terminology and how threats and mitigations are connected. The IRM game's design facilitates a feedback loop where students understand what they have learned, recognise gaps in their understanding and receive guidance on how to improve.

Peer Feedback

The Red vs Blue game framework when played in small groups of students allows for peer feedback where students review each other's strategies and decisions. This approach promotes collaborative learning and critical thinking as students are encouraged to articulate their understanding and challenge their peers' misconceptions. Peer-to-peer feedback has been shown to lead to significant learning gains by fostering focused discussions and deeper engagement with the content (Ambrose et al., 2010). This happens

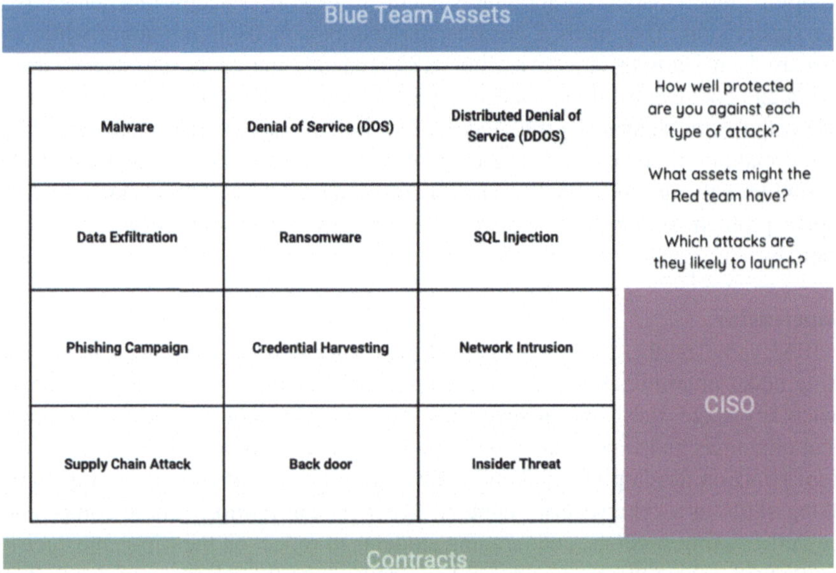

Fig. 12. The strategy board used by a Blue Team to track which attacks they can mitigate against.

intuitively due to the game's competitive nature as attack and defence teams compete against each other.

Reflection and Developing Metacognition

It is important that after gameplay, teachers facilitate debriefing sessions where students reflect on the strategies they used, what worked and what didn't and why. This reflective process helps students consolidate learning and understand the relevance of the skills they have developed to connect their in-game decisions to broader cybersecurity principles. This aligns with best practices in computing pedagogy, emphasising the importance of developing metacognition (Education Endowment Foundation, 2021).

The introduction of a strategy board (see Fig. 12) within the IRM game scaffolds learning by providing students with a clear visual representation of potential attacks and defences. This tool helps students develop a game strategy by understanding their level of protection and identifying potential threats. By reducing cognitive load, the strategy board allows students to focus on strategic decision-making rather than memorising which threats they can launch or need to mitigate. This shared view of the threat landscape aids in discussions during debriefing sessions, enabling students to articulate their approaches and learn from each other's experiences. The strategy board thus enhances the overall feedback process by supporting strategic thinking and collaborative learning, making the game more effective as an educational tool.

6 Discussion

Drawing on this pedagogic case and its intricacies, three discussion themes in relation to learners are identified: Mindset, Skillset and Understanding. The IRM game promotes a growth mindset by encouraging students to learn from failures and persist through challenges. In the game, students face complex cybersecurity scenarios that require them to think critically and adapt their strategies. This iterative nature of learning reflects the real-world environment of cybersecurity, where professionals continuously learn from past experiences and refine their approaches. By embracing challenges and viewing failures as opportunities for growth, students develop resilience and a positive attitude towards learning, which are crucial traits for success in the cybersecurity industry.

Cyber security requires an interdisciplinary approach, combining technical and non-technical skills. The IRM game effectively integrates these skills, demonstrating the importance of both in the field. Students develop technical skills such as threat analysis, risk management, and incident response, while also honing soft skills like communication, teamwork, and strategic planning. The collaborative elements of the game enhance these skills in a realistic context, preparing students for the multifaceted nature of cybersecurity roles. This integration helps students build a comprehensive skillset that is highly valued in the industry.

The unplugged format of the IRM game supports students to develop a more flexible mindset around the skills needed in cyber roles challenging the stereotype that all cyber roles are highly technical. The tangible, hands-on nature of card-based gameplay provides a break from screen-based learning, fostering a more inclusive and varied learning experience making cybersecurity concepts accessible and engaging without relying on technology. Research has shown that unplugged activities can be particularly beneficial in encouraging girls to pursue computing. These activities highlight the creative and collaborative aspects of computer science, counteracting stereotypes that may deter girls from engaging with the subject (Chen et al., 2023). By focusing on problem-solving, teamwork, and strategic thinking without the immediate pressure of technology, the unplugged format creates a supportive and welcoming environment that can boost confidence and interest among female students, helping them see computing as relevant and achievable.

The immersive and iterative nature of the IRM game helps students develop a deep, nuanced understanding of cybersecurity concepts and their real-world applications. By engaging in hands-on, scenario-based learning, students move beyond surface-level engagement to grasp the complexities of the field. Ensuring that the game and its content are inclusive and culturally relevant is vital for engaging all students and making learning more relatable. The IRM game reflects diverse perspectives and experiences, promoting an inclusive learning environment. This culturally relevant pedagogy helps in engaging students from different backgrounds, making cybersecurity accessible and appealing to a broader audience. By showcasing a variety of roles and scenarios, the game highlights that cybersecurity is a field for everyone, counteracting stereotypes and encouraging a more diverse group of students to pursue careers in this area.

The IRM game plays a crucial role in reshaping students' understanding of cybersecurity careers, addressing and dismantling the stereotypical view that all cyber roles are highly technical. By integrating scenarios that highlight the significance of non-technical roles such as risk management, ethical decision-making, and strategic planning, the game emphasises that cybersecurity is a diverse field requiring a wide range of skills. This approach demonstrates that diversity in team roles strengthens overall cybersecurity efforts. The game's design aligns with Gatsby Benchmark 3[11] by addressing the needs of each student through personalised, inclusive learning experiences that cater to various strengths and interests. It also meets Gatsby Benchmark 4 by linking curriculum learning to careers, illustrating the relevance of classroom knowledge to real-world applications. By embedding these principles within the gameplay, the IRM game encourages students to appreciate the value of non-technical roles, fostering a more comprehensive understanding of the cybersecurity landscape and promoting the idea that diverse skills and backgrounds are essential for building robust and effective teams in the field.

While the game cards themselves provide (limited) detail around the assets and attacks it has been suggested that providing more detailed documentation would provide an deeper understanding of the different attacks, technical details on their impact (and mitigations) and possible case studies. This would give teachers more confidence when running sessions and fielding questions from students themselves. It would also be useful to provide links and websites for directed self-study, giving student who are interested in cyber security targeted resources to look at, rather than attempting research using only terms or key words picked up from the game itself.

[11] https://www.gatsby.org.uk/education/focus-areas/good-career-guidance.

7 Conclusion

This chapter presents a case study demonstrating how teachers utilised the IRM game as a pedagogic tool. As evidenced through the chapter, the game encouraged and emphasised the advantage of team-based play styles as key success factor. The students enjoyed the immersive experience that the game provided them with, and positively commented on how the game was providing a strong foundation for future employability with its emphasis on risk management, the handling of cyber security threats, and the pressure of working within a high-stake environment. In this chapter we have highlighted the current gaps in cyber security industry and academia, namely how the historic focus on technical skills and certification have left a significant gap when it comes to areas such as security and risk management, as well as associated professional skills. In response to our collaboration with regional schools we have identified that this has fed into the viewpoint that a lack of technical skills can be a "barrier to entry" for cyber security, with many younger students being unaware of job roles within the cyber security industry beyond "hacker". Industry, academia and teachers have come together and have mapped the game play to existing educational and industry frameworks, as well required cyber security career skills. Importantly as the game was designed with teachers in mind, and used feedback from them at multiple development stages, it aligns with existing GCSE specifications (such as OCR and AQA), and subsequently allows teachers to utilise the game as a lesson which can easily fit into timetabled teaching and curriculum. We have made this game openly available, with ready to print templates, to ensure it is scalable and easy to access and use. We hope that the IRM game will be a source of inspiration, motivation and importantly preparation for a critical industry such as cyber security.

References

Ambrose, S.A., Bridges, M.W., DiPietro, M., Lovett, M.C. Norman, M.K.: How Learning Works: Seven Research-Based Principles for Smart Teaching. Wiley, New York (2010)

Angafor, G.N., Yevseyeva, I., He, Y.: Game-based learning: a review of tabletop exercises for cybersecurity incident response training. Secur. Privacy **3**(6), e126 (2020)

BBC News. Classes cancelled as 'sinister' school cyber-attacks rise (2024). https://www.bbc.co.uk/news/articles/c2vwz4exq4xof. Accessed 18 July 2024

Chen, P., Yang, D., Metwally, A.H.S., et al.: Fostering computational thinking through unplugged activities: a systematic literature review and meta-analysis. Int. J. STEM Educ. **10**(47) (2023). https://doi.org/10.1186/s40594-023-00434-7. Accessed 18 July 2024

Curzon, P., McOwan, P.W., Donohue, J., Wright, S., Mars, D.W.: Teaching of concepts. In: Sentance, S., Barendsen, E., Schulte, C. (eds.) Computer Science Education: Perspectives on Teaching and Learning in School, pp. 91–108. Bloomsbury Publishing, London (2018)

Department for Science, Innovation and Technology. Cyber security skills in the UK labour market (2023). https://assets.publishing.service.gov.uk/media/64be95f0d4051a00145a91ec/Cyber_security_skills_in_the_UK_labour_market_2023.pdf. Accessed 23 May 2024

Education Endowment Foundation (EEF). Special Educational Needs in Mainstream Schools: Evidence Review (2020). https://d2tic4wvo1iusb.cloudfront.net/documents/guidance/EEF_SEND_Evidence_Review.pdf?v=1678717911. Accessed 18 July 2024

Education Endowment Foundation (EEF). Metacognition and Self-Regulation. Teaching & Learning Toolkit (2021). https://educationendowmentfoundation.org.uk/education-evidence/teaching-learning-toolkit/metacognition-and-self-regulation. Accessed 18 July 2024

Goupil, F., Laskov, P., Pekaric, I., Felderer, M., Dürr, A., Thiesse, F.: Towards understanding the skill gap in cybersecurity. In: Proceedings of the 27th ACM Conference on Innovation and Technology in Computer Science Education, vol. 1, pp. 477–483, July 2022

Hallet, J., Shreeve B.: Mapping of cybersecurity games onto CyBOK (2022). https://www.cybok.org/media/downloads/Mapping_cybersec_games_to_CyBOK_report.pdf. Accessed 9 July 2024

National Centre for Computing Education (NCCE). Encouraging Girls into Computer Science Handbook (2021). https://teachcomputing.org/i-belong. Accessed 18 July 2024

Ramezan, C.A.: Examining the cyber skills gap: an analysis of cybersecurity positions by sub-field. J. Inf. Syst. Educ. **34**(1), 94–105 (2023)

Shreeve, B., Gardiner, J., Hallett, J., Humphries, D., Rashid, A.: Decisions & Disruptions 2: Decide Harder. arXiv preprint arXiv:2306.12168 (2023)

Vygotsky, L.S.: Mind in Society: The Development of Higher Psychological Processes. Harvard University Press, Cambridge (1978)

Walkington, H., Dyer, S., Solem, M., Haigh, M., Waddington, S.: A capabilities approach to higher education: geocapabilities and implications for geography curricula. J. Geogr. High. Educ. **42**(1), 7–24 (2018)

Winger, G., Ellis, S., Glover, D.: Bridging the digital gap: teaching cyber strategy and policy through a crisis simulation. Int. Stud. Perspect. **25**(2), 145–163 (2024)

Author Index

P. Legg et al. (Eds.): CSE 2024, LNNS 1213, p. 235, 2024.
https://doi.org/10.1007/978-3-031-77524-6

The manufacturer's authorised representative in the EU is Springer Nature Customer Service Centre GmbH, Europaplatz 3, 69115 Heidelberg, Germany. If you have any concerns regarding our products, please contact ProductSafety@springernature.com

Printed and bound by CPI Group (UK) Ltd, Croydon, CR0 4YY
01/10/2026
02225911-0004